Robert L. Vaught

Set Theory
An Introduction

Second Edition

Birkhäuser
Boston • Basel • Berlin

Robert L. Vaught
Department of Mathematics
University of California
Berkeley, CA 94720
U.S.A.

**The Library of Congress has cataloged the
[hardcover imprint] edition as follows:**

Vaught, Robert L., 1926-
Set theory : an introduction / Robert L. Vaught. – 2nd ed.
p. cm,
Includes bibliographical references and index.
ISBN 0-8176-3697-8
1. Set theory. I. Title.
QA248.V38 1994 95-9544
511.3'22–dc20 CIP

Printed on acid-free paper.

Birkhäuser

ISBN 0-8176-4256-0 SPIN 10843280
ISBN 3-7643-4256-0

Printed and bound by Berryville Graphics, Inc., Berryville, VA
Printed in the United States of America

9 8 7 6 5 4 3 2 1

Dedicated to my wife Marilyn

TABLE OF CONTENTS

PREFACE

By its nature, set theory does not depend on any previous mathematical knowledge. Hence, an individual wanting to read this book can best find out if he is ready to do so by trying to read the first ten or twenty pages of Chapter 1. As a textbook, the book can serve for a course at the junior or senior level. If a course covers only some of the chapters, the author hopes that the student will read the rest himself in the next year or two. Set theory has always been a subject which people find pleasant to study at least partly by themselves.

Chapters 1-7, or perhaps 1-8, present the core of the subject. (Chapter 8 is a short, easy discussion of the axiom of regularity). Even a hurried course should try to cover most of this core (of which more is said below). Chapter 9 presents the logic needed for a fully axiomatic set theory and especially for independence or consistency results. Chapter 10 gives von Neumann's proof of the relative consistency of the regularity axiom and three similar related results. Von Neumann's 'inner model' proof is easy to grasp and yet it prepares one for the famous and more difficult work of Gödel and Cohen, which are the main topics of any book or course in set theory at the next level. Chapter 9 might be slightly easier for someone who has already studied logic, but it is written to be understandable by a reader with no background in logic. Actually, some of the logic given in Chapter 9 is not covered in most first year logic courses (and most of what they do cover is not needed in Chapter 9 or 10). After Chapters 1-8, the thing most required for further work in set theory (and so the thing next to be included in a longer course) is the material of Chapter 9 (and Chapter 10). The last Chapter, 11, returns to 'straight' set theory and can be read after Chapter 7. Its first part adds to the earlier cardinal arithmetic, and its second part to the earlier ordinal arithmetic. Most people find these topics attractive and easy, and will indeed read them by themselves if a course does not cover them.

For many years, the widely used introductory books on set theory all presented intuitive set theory. For the past two or three decades, the exact opposite has been true: all such books have given axiomatic set theory. But for the student, the trivial and irritating business of fooling around, as he begins to learn set theory, with axioms (saying for example that $\{x,y\}$ exists!) discourages him from grasping the main, beautiful facts about infinite unions, cardinals, etc., which should be a joy.

Therefore, we shall work in intuitive set theory in the first five of the seven main chapters. The axioms are discussed in the very short Chapter 6. By that time, many of the special features of the axiomatic business will be seen by the student to be trivial, as they should be. At the end of Chapter 6 the reader has all of Chapters 1-5 behind him *axiomatically*. In Chapter 7 (on well-orderings) we now work from the axioms, but the reader sees at once that there is practically no difference between working intuitively and working axiomatically.

Two other pedagogical devices are used to increase the reader's speed in getting the main ideas – the first (which the author learned from Azriel Levy when he was teaching in Berkeley) is this: Cardinals, order types, etc., are not defined (in some ad hoc way) until Chapter 7; but, in Chapter 2, we just 'grab' them, as Cantor did. The other device imitates the famous book (or books) of Hausdorff [F1914 and F1927 (English edition 1957)] in putting off the serious study of well-orderings as long as possible – in fact until Chapter 7. (Even in Cantor's work, some ideas are less natural and easy then others!) As a side-effect, well-ordering is studied (in Chapter 7) while working axiomatically; and it is just possible that this subject (particularly definition by induction) is one of the few more easily grasped working axiomatically than intuitively.

The author is indebted to his own teacher in set theory and logic, Alfred Tarski, for many things in this book. Some recent students have suggested various shorter proofs, which have been gratefully used. The author is very grateful to Shaughan Lavine, who prepared the index and also assisted with the proofreading, making many corrections and improvements.

INTRODUCTION

Set theory has two overlapping aspects. In one, it is a branch of mathematics, like algebra or differential geometry, with its own special subject matter. In its other aspect, set theory is not a branch of mathematics but the very root of mathematics from which all branches of mathematics rise. (In this picture only logic lies still below set theory. Together they are often called the 'foundations of mathematics.')

This Introduction contains some remarks about the (early) history of set theory. The book proper, the mathematics, begins with Chapter 1 and does not depend on the Introduction. The remarks below should be read (rapidly for pleasure) now or perhaps after reading much of the book, or both.

Set theory, in its aspect as one branch of mathematics, is devoted particularly to the theory of infinite cardinal and ordinal numbers. Perhaps not even relativity theory can be said to have sprung so completely from the mind of one man as did set theory (in this aspect)! That man was Georg Cantor. Cantor lived from 1845 to 1918, his main publications appearing between 1874 and 1897. (For reference to Cantor's papers see the bibliography of Fraenkel [1960].) Cantor was also one of the founders of point set topology which in turn had arisen in a study of trigonometric series. Cantor's set theory stands as one of the great creations of mathematics. David Hilbert is widely considered to be the leading mathematician of the last hundred years. Replying to some who thought that the paradoxes (see below) might destroy Cantor's theory of sets, Hilbert spoke of "a paradise created by Cantor from which nobody shall ever expel us" (cf. Fraenkel [1961], p. 240). Reader: Note what lies ahead for you!

The early story of the 'foundations aspect' of set theory goes back farther, and it is not concentrated in the work of one man. Actually, in the years 1800-1930 the entire mathematics went through a major change (at the very least, of *style*) into what every subject calls today its 'modern approach.' Set theory and logic played a key role in this change, but that role is nevertheless sometimes over-estimated. In fact, every branch of mathematics was going through the same convulsions. For example, in algebra the (central modern) notion of isomorphism began to appear by 1830 or earlier in the work of Galois and others, but the simple, general modern concept is not exactly present until perhaps 1910 (Steinitz on fields) or even 1920! One might think nothing special was happening here, as mathematics is always evolving. But B.L. von der Waerden's *Modern Algebra* [F1950], a graduate-level algebra book, appeared first in 1930, and is still widely used as a textbook today, fifty years later! During 1850-1930 the serious changes in style always going on would have made such a thing almost impossible. An essential feature of the modern approach is just that there is in a sense an *end of the line* in style and rigor. In most fields that was reached within ten years of 1930!

The early developments in (the foundations aspect of) set theory were closely connected with the 'convulsions' in *analysis* in the 19th century. In a reversal of the lack of rigor in the 18th century, the modern, rigorous approach to analysis began to appear in the early 1800's. By 1830, Cauchy and others were able to use *almost* the modern style in defining limits or continuity. One man ahead of his time went even further than Cauchy towards our modern analysis, namely the priest, B. Bolzano (1781-1848) (see [F1810], [F1837], and [F1851])*. Bolzano began, in particular, to use the notion of (arbitrary) *set* much more in analysis. In the same period the closely related notion of arbitrary *function* was emerging, e.g., in the work of Fourier (1810), Dirichlet (1840), and Riemann (1826-1866). (One date as in "Fourier (1810)," refers to the time of a key publication). Bolzano is indeed the only person ever proposed as a predecessor for Cantor (and was so even by Cantor himself). Bolzano had begun (but only begun) to study the notion 'A and B can be put in one-to-one correspondence' – the keystone of Cantor's theory.

By 1861, K. Weierstrass was able to give almost our idea of a 'modern' course in real variables! A decade or so later, Dedekind and Cantor created what are still the two main methods for constructing the reals from the rationals. This was a needed step in reducing all of mathematics to set theory. Richard Dedekind (1831-1916), an older mentor and longtime friend of Cantor's, contributed much to the set-theoretical study of the natural numbers. His famous book *Was sind und was sollen die Zahlen* [F1888] is full of passages which were new then but are now spoken in every beginning course in real variables or modern algebra. G. Peano (1895) is also famous for his contributions to the same subject.

Ernst Zermelo [F1904, 1908, 1908a, 1909, 1913] seems to have been the first to take each natural number to be a certain set. He also was the first to know (although in an odd way) how to get along with only $\in$ and without the ordered couple† (though later Norbert Wiener in 1915 pointed out the much more elegant

---

* References like "Bolzano [F1851]" are explained at the beginning of the Bibliography.

† Roughly, Zermelo observed he could get along using only functions $f: A \to B$ where A and B are disjoint; and that such an f can be taken to be a suitable set of doubletons $\{a,b\}$. (The author learned this from Gregory Moore.)

fact that one can simply *define* a reasonable notion of ordered couple using only $\in$.) These two steps taken by Zermelo were among the last needed to show that one could do the entire mathematics in a set theory using only the notion 'belongs to.' Clearly these steps were not unrelated to Zermelo's most famous contribution, made in 1908, when he gave the first axiom system for set theory (and hence for all of mathematics). The most frequently used axioms today are nearly just his! Nevertheless, Zermelo's axiomatic work was lacking in one respect, as he did not realize that an axiomatic system cannot be fully understood until its underlying logic is fully understood.

Earlier, at exactly the same time as Cantor, there was a man, of perhaps the same brilliance as Cantor, who leaped ahead fifty years in the subject of *logic*. That man was Gottlob Frege. (His key publications were in 1879-1903). Some of his contributions will be mentioned a bit later on.

In the years 1897-1902, paradoxes (that is, contradictions) were discovered by Bertrand Russell and others using what perhaps *might* be taken as acceptable set-theoretical axioms. (Frege, almost alone, had actually assumed these axioms!) For many decades, the paradoxes were considered to be the central feature of set theory or at least of its foundations. For example, many books seemed to imply that the whole purpose of the axiomatic approach in set theory was to deal with the paradoxes (awkward for Euclid!) In recent times, the paradoxes have been assigned a lesser importance, though certainly a great one. In fact, Cantor himself knew that it was necessary to distinguish between 'ordinary' sets and very large, 'bad' collections. Obviously, Zermelo had very good reasons for studying the axiomatization of set theory (and the whole of mathematics!) – even without the paradoxes. But the paradoxes have certainly played a central role in mathematical philosophy for decades. Another controversy has been over the axiom of choice – which was made famous by Zermelo in 1904 when he derived from it the well-ordering principle. Both the paradoxes and the axiom of choice will be discussed (mathematically) in later chapters, where some brief remarks will be made on their history and, in general, on the history of set theory after 1900.

Let us return to the history of a fully correct axiomatization of set theory (which we shall follow up to 1930). Frege had provided just the understanding of logic and logistic systems needed for a fully correct axiomatization. (A logistic system is a deductive system in which the notion of what is a proof is so clear it can be decided by a machine.) Unfortunately, Frege's work was not widely known even by 1900. However, at almost the same time as Zermelo's axiomatization, another very different axiomatization of set theory was carried out by Bertrand Russell and Alfred North Whitehead in their famous three volume work, *Principia Mathematica.* From the point of view of a working mathematician, Zermelo's axioms were (and are) tremendously better than the awkward system of the *Principia.* However, Russell and Whitehead knew and appreciated Frege's work in logic. They included formal logic in their work and, although possibly with some flaws, their work was a logistic system (for set theory and the whole of mathematics). It is thus clear why this awkward work was so acclaimed and so influential.

There remained the (actually very easy) problem of getting a system both workable and fully logistic. In [F1923], Thoralf Skolem succeeded, say, 95% in simply making the working system of Zermelo into a logistic one. He proposed that Zermelo's 'schemas' (which caused the trouble) be replaced by the infinite set of their '∈-instances' (exactly as is often done today). But, alas, it seems unlikely that Skolem grasped then the idea of a logistic system for logic. (In his other papers of the time, he always considers that the only meaning of 'logically valid' is 'holds in all models.') Nevertheless, it seems that in 1923, to the advanced logicians in the famous 'Hilbert group' in Germany and the famous 'Polish group,' Skolem's paper must have suggested the full logistic system, just as it does today.

In the very same paper, Skolem shared at least partly with Adolf Fraenkel [L1922] and Dmitri Mirimanoff [L1917] the (independent) discovery of the principal missing axiom in Zermelo's system, needed to carry out Cantor's original work – the axiom of replacement. (Fraenkel's influence in the matter was greatest and the axiom is often called 'Fraenkel's axiom.')

There is one other axiom usually included today, namely, the axiom of regularity. This axiom concerns 'partial universes' which are loosely related to Russell's 'types.' The partial universes and the axiom of regularity were first considered by Mirimanoff [L1917] (and also in Skolem [F1923]). They were given a definitive study by Johan (later John) von Neumann in a group of papers on set theory [F1923-1928].

In the same group of papers, the young von Neumann, who was to become perhaps the greatest mathematician of the first half of the twentieth century, published the first flawless axiomatization of set theory. In fact, it had its own awkwardness and has rarely been used! A workable modification of it was given by P. Bernays [F1937-1954]. Notice that the two systems of set theory (which do not differ very much) most used today (Zermelo-Fraenkel and von Neumann-Bernays) were achieving exactly their present form in just the years when van der Waerden's *Modern Algebra* appeared!

1. SETS AND RELATIONS AND OPERATIONS AMONG THEM

The mathematical content of Chapter 1 is easy, so the chapter should be read rather rapidly.

Before we begin to study sets, a brief remark about logic will be made. The reader is probably already familiar with at least some of the following logical symbols or abbreviations:

$\sim$	for	*not*, *or it is not the case that*.
$\vee$	for	*or*.
$\wedge$	for	*and*.
_____ $\rightarrow$	for	*if* _____ *then* ....
_____ $\leftrightarrow$	for	_____ *if and only if* .... (Also: 'iff'.)
$\forall x$ _____	for	*for all x*, _____ .
$\exists x$ _____	for	*for some x*, _____ or *there exists an x such that* _____ .
$\exists ! x$ _____	for	*there exists exactly one x such that* _____ .
ιx _____	for	*the unique x such that* _____ .
$x = y$	for	*x equals y*, or *x and y are identical*.

In theoretical discussions about the language we use – for example in Chapters 6 and 9 – it is very useful to suppose, or pretend, that we write in a purely symbolic language.

On the other hand, in actually doing mathematics it is desirable to write in pure English (or whatever ordinary language), not using any logical symbols (except $=$). The reason is that in the ordinary language one has a rich tradition of conventions for how to write grammatically and even for how to argue. On the other hand, in 'popular symbolic logic' there is no convention for dealing (as in a proof) with more than one sentence at a time. A *single sentence* in symbolic logic is readable, and hence logical symbols are occasionally used in the statement of a theorem or definition. If the reader will look at any current journal in mathematics (even on logic) he will find virtually no use of logical abbreviations!

Even though they will not be used much, the logical symbols listed above are helpful in understanding ordinary (English) mathematical usage. For example, we realize that 'for all *x*,' 'for any *x*,' 'for every *x*' and 'for each *x*' all are just the same (namely, '$\forall x$'). Moreover, our attention is called to the great importance in mathematics of the English passages above (corresponding to the symbols). In particular, most people need to overcome some odd initial reluctance to use themselves the phrase 'the unique *x* such that' – which one must do throughout modern mathematics!

§ 1. Set algebra and the set-builder

If we have conceived of some things (of any kind), then we can conceive of sets of these things – and sets of sets, etc., etc. We may have started with some things which are not sets themselves ('non-sets,' 'atoms,' or in German, 'Urelementen'). In one natural form of set theory we make just these assumptions, leaving undecided whether there are a small number or a large number of non-sets. In the most popular set theories today, and in this book, we do make an assumption about the non-sets, namely, that there are not any at all (among the things we choose to consider). Speaking roughly, here are some reasons for making this assumption:

As we will soon see precisely, there is nevertheless an empty set, $\emptyset$; also the set $\{\emptyset\}$ whose only member is $\emptyset$; the set $\{\emptyset,\{\emptyset\}\}$; the infinite list of sets $\emptyset$, $\{\emptyset\}$, $\{\{\emptyset\}\}$, $\{\{\{\emptyset\}\}\}$, etc.; the infinite set containing all of these latter, and so on and so on. One (interesting but not decisive) argument for considering no non-sets (and thus as is said only 'pure' sets) goes as follows: In mathematics we are only concerned with form, and thus with, for example, a group only up to isomorphism. Every group obtained starting with non-sets is isomorphic to one in our world of pure sets; so nothing is lost, perhaps, in considering only pure sets.

In fact, there are times when set theory allowing non-sets is useful, and this theory is not completely covered by our theory of pure sets. But, in practice, it is usually very easy to develop any needed part of set theory with non-sets, if one already knows pure set theory. Also, allowing non-sets causes a lot of trivial difficulties which obscure the things which really matter. So as a practical matter, it is convenient to consider, as we do here, only pure sets (henceforth called plain 'sets.') We may still give informal examples involving non-sets.

Henceforth, letters '*x*', '*y*', '*A*', '*B*', etc., etc., all range over sets (or really, 'pure sets'). Hence, it is optional, for example, whether one says 'for any *A*' or 'for any set *A*.'

We understand as in ordinary English the statement '*x* belongs to *A*' or the synonymous 'x is a member of A', which we abbreviate by: $x \in A$.

Suppose C is the set of all American citizens and D is the set of all American citizens less than twelve feet tall. Then C and D have exactly the same members but have been defined in different ways. Ordinary usage does not take a definitive stand as to whether C and D should be considered to be identical. In mathematical usage, ambiguities are not tolerated. If both usages seems interesting, mathematics just studies both, using two names! In mathematical usage it is generally agreed to use the words 'set' and 'belongs to' in such a way that the sets C and D above *are* considered identical. This convention is expressed systematically in what is called the

Axiom of Extensionality. *If, for any x, $x \in A$ if and only if $x \in B$, then $A = B$.*

(One says '*extensional*' because the different *intensions*, as, say, in our descriptions above of C and D, are being disregarded.) Note that we will mention *some* axioms in the *non-axiomatic* development of Chapters 1-5 (as was always done historically). But in non-axiomatic or intuitive set theory one does not work *only* from the axioms mentioned but allows other intuitions to be used.

We say that A is a *subset* of B, or A *is included in* B, or, in symbols, $A \subseteq B$ if every member of A is a member of B. It follows at once that, in general: $A \subseteq A$ (inclusion is 'reflexive'); if $A \subseteq B$ and $B \subseteq C$ then $A \subseteq C$ (inclusion is 'transitive'). Moreover, the Extensionality Axiom can now be written: if $A \subseteq B$ and $B \subseteq A$ then $A = B$ (inclusion is 'antisymmetric').

Once in a while, instead of discussing sets (which are 'somewhere else'!) we discuss our own language, say, chalkmarks on the blackboard (which are 'in this room'). All expressions in the language of mathematics can be divided (in an extremely important way) into three classes: (1) *asserting* expressions; (2) *naming* expressions; and (3) 'neither of these': The reader will at once be able to classify in this way the following expressions: $x +$; $u + 2$; $x < y$; the unique y such that $x + y = 3$; the set of all x such that $x < 2$; $< x + 3$; $x < y$ or $x < z$; and $\int_1^2 x^2 dx$. Also (using logical symbols) $x < \iota y\,(y^3 = 2)$; $\forall zz$; $\exists!x$ (x is the Queen of England). (The answers are: 3, 2, 1, 2, 2, 3, 1, 2, 1, 3, 1.)

A key notion of set theory is that of forming *the set of all x such that* ..., which is abbreviated $\{x: \ldots\}$. (The *symbols* $\{-: ---\}$ are sometimes called the set-builder. Note to reader: Always read $\{x: \ldots\}$ in full as "the set of all x such that ...".) For example, one can form $\{x: x = 1 \text{ or } x = 2\}$ and $\{x: x \text{ is a real number}\}$, etc.

Clearly, in the expression "$\{x: \ldots\}$", the dots are always to be replaced by an *asserting expression*. (Thus this grammatical notion is just what is needed here!). An extension of this usage is $\{---: \ldots\}$ as in $\{n^3 + n: n \text{ is a positive integer}\}$ which is read: the set of all $n^3 + n$ such that n is a positive integer. In $\{---: \ldots\}$, the '...' is still to be asserting, but the '$---$' is clearly to be a naming expression. These two set-builder notions do not have to be taken as primitive (or governed by the English meaning), but can be introduced in a simple way *by definition*, as follows:

Definition 1.1.

(a) $\{x: \mathcal{P}x\}$ = *the unique A such that for any x, $x \in A$ if and only if $\mathcal{P}x$.*
(b) $\{\mathcal{A}_x: \mathcal{P}x\} = \{u$: *for some x*, $u = \mathcal{A}_x$ *and* $\mathcal{P}x\}$.

In Definition 1.1 we are using in a special, familiar way, the capital script letters '$\mathcal{P}$', '$\mathcal{Q}$', etc. and also in a related but different way the capital script letters '$\mathcal{A}$', '$\mathcal{B}$', etc. By properly using these letters, we can avoid the awkward and unclear '. . .' and '– – –' which we used above. Sometimes '$\mathcal{P}$' (for example) has, say, *two* places instead of one as in 1.1, and then one writes '$\mathcal{P}ab$', '$\mathcal{P}xx$', etc. Of course, for example, '$\mathcal{P}x$' is an *asserting expression*; while in 1.1(b) (and always) '$\mathcal{A}_x$' is a *naming expression.*

The letters '$\mathcal{P}$', '$\mathcal{Q}$' are called *predicate* (or sometimes, *class*) *variables*; while '$\mathcal{A}$', '$\mathcal{B}$', etc. are called *operator variables*. Our old letters 'x', 'y', 'x'', · · ·, 'A', 'B', · · · are commonly called *ordinary variables* or, simply, *variables.*

As usual, if we accept as valid a statement involving, say, the predicate variable '$\mathcal{P}$', like 1.1(a) above, then we also accept as valid each *instance* of our statement, that is, each statement obtained from the original one by substituting for '$\mathcal{P}$' a particular asserting expression. (For example, in 1.1(a) one might replace $\mathcal{P}x$ throughout by, say, $\exists u(x \in u \wedge u \in B)$. Similarly (in forming an instance) one can replace, e.g., '$\mathcal{A}$' in a given statement by a particular naming expression (say, replace $\mathcal{A}_x$ by $x^2 + xy$).

We consider now the actual meaning of $\{x: \mathcal{P}x\}$, that is: the unique set A such that for any x, $x \in A$ iff $\mathcal{P}x\}$. Clearly, by Extensionality, *there is always at most one such A*. The famous 'paradoxes' of set theory (cf. §2) involve cases of $\mathcal{P}x$ where there is no such A, i.e., $\{x: \mathcal{P}x\}$ does not exist! In *intuitive* set theory we adopt the (dubious) attitude that we will assume that $\{x: \mathcal{P}x\}$ does exist in all cases that seem natural (and hence all that we shall consider without comment); and we will imagine that our good sense will keep us from trying to form $\{x: \mathcal{P}x\}$ in any 'bad' case!

We will now discuss much more straightforward things, namely, the so-called Boolean operations on sets. (These were first studied by George Boole in 1847 and by A. de Morgan in the same period.) We define:

$A \cup B = \{x: x \in A$ or $x \in B\}$, called the *union of A* and B.

$A \cap B = \{x: x \in A$ and $x \in B\}$, the *intersection* of A and B.

$A - B = \{x: x \in A$ and $x \notin B\}$, A *minus* B.

$A \ominus B = (A - B) \cup (B - A)$, the *symmetric difference* of A and B.

$\emptyset = \{x: x \neq x\}$, the *empty set* (clearly by Extensionality the unique set with no members).

$X - A$ is also written $\tilde{A}^{(X)}$ (and called: the *complement of A with respect to*

X). Often we are in an extended discussion in which all sets A, B, C, etc., being considered are subsets of a particular set X (a 'temporary universe'). For example, X might be the set of real numbers or some other 'space'. Then we write $\tilde{A}^{(X)}$ *simply as* $\tilde{A}$ (and call it simply the *complement of* A). We shall soon see (in §2) that there is no absolute universe (set of all sets). It follows that there is no absolute complement of any set!

Picturing A and B as regions in the plane, the Boolean Operations appear as in the parts of Figure 1 (called *Venn diagrams*). Each shaded area represents the set resulting from the indicated Boolean operation.

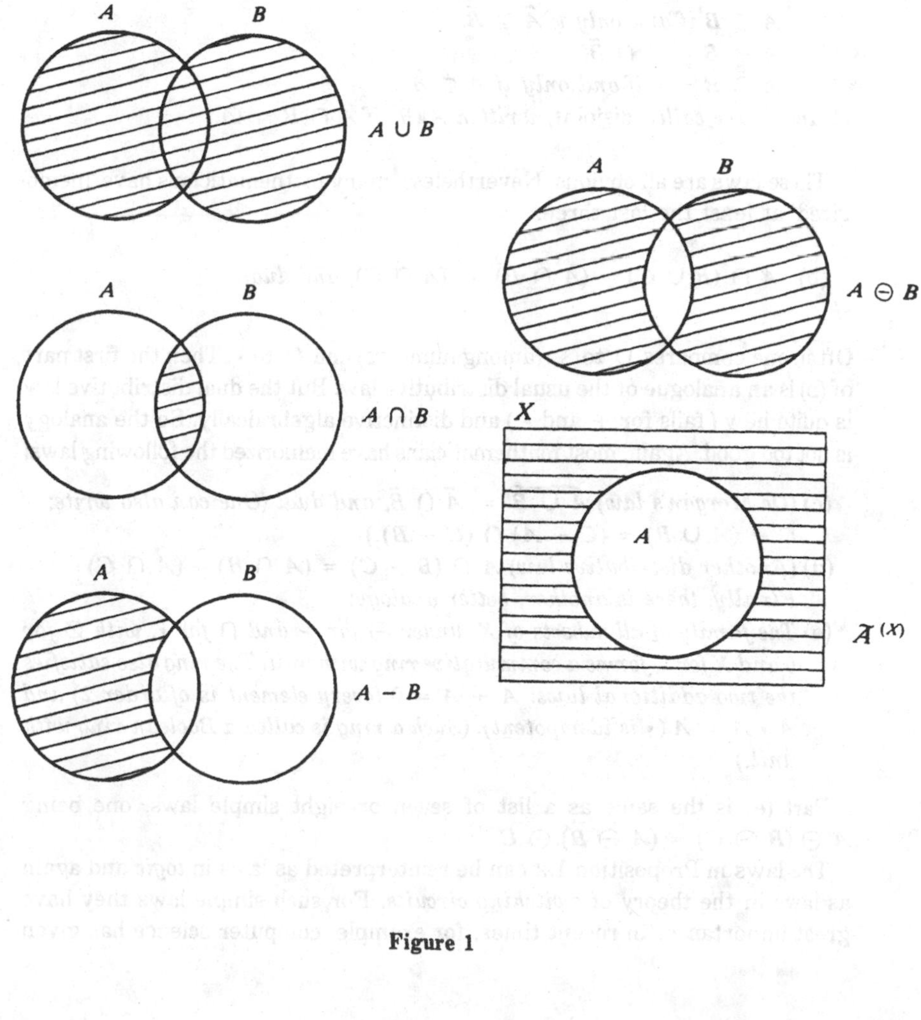

Figure 1

Notice, in the definitions of $A \cup B$, $A \cap B$, etc., above, the close parallel between the Boolean set operations $\sim$, $\cup$, $\cap$ and the logical symbolic $\sim$ (not), $\vee$, $\wedge$. Indeed even the common *symbols* have been chosen so that '$\cup$' resembles '$\vee$', etc.!

We now list some of the laws (all very easily proved) regarding these operations on sets. If a law involves only $\sim$, $\cup$, $\cap$, $\emptyset$, X (not $\ominus$, etc.), its *dual* is obtained simply by replacing these, respectively, by $\sim$, $\cap$, $\cup$, X, $\emptyset$.

Proposition 1.2. *Assume $A, B, C \subseteq X$ and write $\tilde{U}$ for $X - U$.*

(a) $A \cup B = B \cup A$ *and dual.*

$A \cup (B \cup C) = (A \cup B) \cup C$ *and dual.*

$A \cup \emptyset = A$ *and dual* ($A \cap X = A$).

$A \cap \emptyset = \emptyset$ *and dual.*

$\tilde{\tilde{A}} = A.$

$A \subseteq B$ *if and only if* $\tilde{A} \supseteq \tilde{B}$.

$A - B = A \cap \tilde{B}$.

$A \cap B = \emptyset$ *if and only if* $A \subseteq \tilde{B}$.

A and B are called disjoint, *written* $A \between B$, *if* $A \cap B = \emptyset$.

These laws are all obvious. Nevertheless, many mathematicians have 'memorized' at least the last three.

(b) $A \cap (B \cup C) = (A \cap B) \cup (A \cap C)$, *and dual.*

Often one compares $\cup$ to $+$ (among numbers) and $\cap$ to $\cdot$. Then the first part of (b) is an analogue of the usual distributive law. But the dual distributive law is quite new (fails for $+$ and $\cdot$) and distinctive algebraically. So the analogy is not too good. Again, most mathematicians have memorized the following laws:

(c) (*De Morgan's law*) $\widetilde{A \cup B} = \tilde{A} \cap \tilde{B}$, *and dual* (*One can also write*: $C - (A \cup B) = (C - A) \cap (C - B)$.)

(d) (*Another distributive law*) $A \cap (B - C) = (A \cap B) - (A \cap C)$

Finally, there is another, better analogy:

(e) *The family of all subsets of X, under $\ominus$ for $+$ and $\cap$ for $\cdot$, with $\emptyset$ for 0 and X for 1, forms a commutative ring with unit. The ring also satisfies the two additional laws: $A + A = 0$ (every element is of order 2) and $A \cdot A = A$ ($\cdot$ is idempotent).* (*Such a ring is called a Boolean ring with unit.*)

Part (e) is the same as a list of seven or eight simple laws, one being $A \ominus (B \ominus C) = (A \ominus B) \ominus C$.

The laws in Proposition 1.2 can be reinterpreted as laws in *logic* and again as laws in the theory of *switching circuits*. For such simple laws they have great importance. In recent times, for example, computer science has given rise to various new problems like this: How can one 'simplify' expressions like

$$\sim (A \cap (\sim B \cup C)) \cup (A \ominus B)$$

as fast as possible?

Problems

1. State and prove the dual law in 1.2(b).
2. Prove 1.2(c) (i.e, all three laws stated there.)
3. Prove 1.2(d).
4. Show $\ominus$ is associative (annoyingly long).
5. Prove the following strange form of cancellation (the simpler forms are not valid):

   If $A \cap C = B \cap C$ and $A \cup C = B \cup C$ then $A = B$.

Note: Problem 5 illustrates some routine steps in writing proofs. Begin: "Suppose $A \cap C = B \cap C$ and $A \cup C = B \cup C$." Now you want to show $A = B$. This is usually best divided into $A \subseteq B$ and $B \subseteq A$. So first you want to show $A \subseteq B$, i.e., for any x, if $x \in A$ then $x \in B$. So next write: 'Let $x \in A$'; and of course try to argue that $x \in B$. These routine steps are taken to allow one to argue at the simplest level. Their value cannot be overestimated.

6. (On language) Classify as (1) an asserting expression, or (2) a naming expression, or (3) neither:

$\int_a^b x^2 d$; the unique x such that; $\exists x (x > y)$; $(A \cup B) \cup C$; $A \cup B \in C$; $\{x: x < 2\}$.

§2. Russell's paradox

Around the turn of the century, the famous antinomies of Burali-Forti (who, as Halmos [1960] says, was one man!), Cantor and Bertrand Russell were discovered. Russell's was not the first but it is very much simpler than the others. These antinomies all involved the unrestricted use of $\{x: \mathcal{P}x\}$ (i.e., with no restriction placed on the asserting expressions taken for '$\mathcal{P}x$'). Russell's is as follows. We take '$x \notin x$' for '$\mathcal{P}x$'. If we could always form $\{x: \mathcal{P}x\}$, then one could here, getting a set $A = \{x: x \notin x\}$. But clearly: if $A \in A$ then $A \notin A$; and if $A \notin A$ then $A \in A$! Thus neither $A \in A$ nor $A \notin A$ can hold, a contradiction. Instead of regarding this as a paradox, we can just say that we have proved the following theorem:

Theorem 2.1. $\{x: x \notin x\}$ *does not exist.*

(It is natural to ask at this point: *are* there any sets z such that $z \in z$? It turns out that in our axiomatic set theory this question will remain unresolved until the last, and very little used, axiom, the Regularity Axiom, is added – see Chapter 8. But notice that in fact *Russell's argument above is completely indifferent to the answer to this question.*)

By 2.1, the set $\{x: \mathcal{P}x\}$ cannot always be found. We nevertheless try to accept a lot of cases of "$\{x: \mathcal{P}x\}$ exists." The two most sweeping axioms in set theory (axiomatic or intuitive) are of this type. In our intuitive set theory, we now accept these two axioms, namely:

2.2.

(a) **Separation** (*or 'Aussonderungs'*) **Axiom** (*Zermelo*): $\{x: x \in B$ *and* $\mathcal{P}x\}$ *always exists.*

(Just as a trivial matter of notation, this set is also written $\{x \in B: \mathcal{P}x\}$, read "the set of all x belonging to B such that $\mathcal{P}x$.)

(b) **Replacement Axiom** (*Fraenkel and Skolem*): $\{\mathcal{a}_x: x \in B\}$ *always exists.*

('Aussonderung' means 'separating out' (like cutting a piece from a pie), which is clearly just right. There is no good translation in English and as a result Axiom (a) has many other names in English: 'specification,' 'subset formation,' etc., etc.).

Using 2.2(a) we can now employ Russell's argument to show that, as stated earlier:

Theorem 2.3. *There is no universe set, i.e., for any set C there exists t such that $t \notin C$.*

Proof. If everything belongs to C, then the set $\{x \in C: x \notin x\}$, which can be formed by the Separation Axiom, would just be $\{x: x \notin x\}$, which can't exist by 2.1.

We shall try to deal with the paradox by, roughly speaking, forming $\{x: \mathcal{P}x\}$ only when the set is not much 'bigger' than some sets we already have. In 2.2 (a) the new set is a subset of an old set so certainly not 'bigger'. In 2.2(b) the new set is not bigger than the old in the sense (rough, new) that it is the image of the old set by a mapping. Of course, 2.3 certainly warns us that bigness is bad! But the idea 'big' remains vague and our remarks about 'bigness' are to be taken only as heuristic.

Problem

1. Improve 2.3 by obtaining (given any C) a set t such that $t \notin C$ and also (*a*) t is a subset of C, and (*b*) t is a specific ('definable') set. Hint: Take $t = \{x: \dots\}$ for a suitable ... (this automatically makes t definable), and expect to use 2.2(a) and an argument like that in 2.1.

§3. Infinite unions and intersections

The operations of taking intersections and unions of two sets are easily extended to any number of sets, even infinitely many. We put the *union* $\bigcup_{i \in I} \mathcal{A}_i = \{x\colon \text{for some } i \in I, x \in \mathcal{A}_i\}$. Similarly, if $I \neq 0$, we put the *intersection* $\bigcap_{i \in I} \mathcal{A}_i = \{x\colon \text{for all } i \in I, x \in \mathcal{A}_i\}$. If $I = \emptyset$ the same definition of $\bigcap_{i \in I} \mathcal{A}_i$ would yield the absolute universe, contrary to 2.3; so $\bigcap_{i \in I} \mathcal{A}_i$ is not defined when $I = \emptyset$. However, if we have (as with complements) a 'temporary universe' X, then we define $\bigcap_{i \in I}^{X} \mathcal{A}_i = \{x \in X\colon \text{for any } i \in I, x \in \mathcal{A}_i\}$ without exception (and the upper 'X' is usually omitted). (This is very convenient when X is some kind of space being studied, as in topology or measure theory).

We can also write, for example, $\bigcup_{A \in K} A$, as this means the above when $\mathcal{A}_A = A$ for all A. The laws regarding $\bigcup$ and $\bigcap$ are again easy to prove and very often used:

Proposition 3.1. *Assume $\mathcal{A}_i \subseteq X$ for all $i \in I$, and write $\tilde{U}$ for $X - U$ and take any intersection of the form $\bigcap_{i \in \emptyset} B_i$* to be X. Then:

(a) (*de Morgan's law*) $\widetilde{\bigcup_{i \in I} \mathcal{A}_i} = \bigcap_{i \in I} \tilde{\mathcal{A}}_i$ *and* $\widetilde{\bigcap_{i \in I} \mathcal{A}_i} = \bigcup_{i \in I} \tilde{\mathcal{A}}_i$.

(b) (*distributive laws*) $B \cap \bigcup_{i \in I} \mathcal{A}_i = \bigcup_{i \in I} (B \cap \mathcal{A}_i)$ *and*
$B \cup \bigcap_{i \in I} \mathcal{A}_i = \bigcap_{i \in I} (B \cup \mathcal{A}_i)$.

A more general distributive law is given in §8. There are also some obvious infinite commutative and associative laws (which are more obvious if not stated!).

Like the set-builder notation '$\{ - : - - - \}$', the *symbol* '$\bigcup$' in '$\bigcup_{i \in I} \mathcal{A}_i$' is what is called a variable-binding operator. A familiar example is the integral sign $\int$ in the expression '$\int_a^b (x^2 + y)\,dx$', which *binds* 'x'. In each meaningful expression (i.e., asserting or naming), each occurrence of an ordinary variable (like 'x' or 'C') is classified as *bound* (or what is the same, *dummy*) or not bound (also called *free*). This notion (to be clarified below by examples) is perhaps the most important grammatical distinction after the one distinguishing asserting and naming expressions. As with the latter, the reader will find he already knows how to make these distinctions.

Examples. In

$$x \in \{y\colon y > 2\}$$

'x' occurs free and both occurrences of y are bound. (The first occurrence of 'y' is sometimes called *binding* as well as bound.) In

$$u + v > \int_a^2 (x^2 + y)\,dx$$

'a', 'u', 'v', and 'y' occur free, and both occurrences of 'x' are bound. In

There exists an x such that $x^2 > 5$

both occurrences of 'x' are bound. In

$$\{x: x \in A \text{ or } x \in B\}$$

and also in

the unique x such that $x > 0$ and $x = 2$

all occurrences of 'x' are bound. In

$$x < 2 \wedge u \in \{x: x \text{ is a prime}\}$$

the first occurrence of 'x' is free, while the others are bound.

For other examples, see Problem 4 below.

Incidentally, there is another notation for union and intersection (having no bound variables). For any family K of sets, we write $\bigcup(K)$ for $\bigcup_{A \in K} A$ – and similarly for $\bigcap(K)$. (In calculus there are similarly the two notations $\int_a^b f(x)dx$ and $\int_a^b f$.) The advantage of the (bound variable) notation $\bigcup_{i \in I} \mathcal{A}_i$ appears, e.g., in 3.1(b) above, where $\bigcup_{i \in I}(\mathcal{B} \cap \mathcal{A}_i)$ is quite awkward to express in the $\bigcup(K)$-notation.

Problems

1. Prove the first law in 3.1(a).
2. Prove the second law in 3.1(b).
3. In the plane, let $C_r = \{P: d(\mathcal{O}, \mathcal{P}) < r\}$ and $C_r^{\cdot} = \{P: d(\mathcal{O}, \mathcal{P}) \leq r\}$. What is $\bigcap_{n=1}^{\infty} C_{1+\frac{1}{n}}$?

(Note the switch from open circles to a closed circle. Obviously, here, unions and intersections are closely related to limits.)

(In a few problems, like 3, we assume knowledge outside of the main development to get informal examples.)

4. (On language) In each expression classify each occurrence of any variable as free or bound.

   (a) For all y, $y < x$.
   (b) $y = \int_1^u (x^2 + y)dy$.
   (c) $\Sigma_{i=1}^n (i^2 + 1)$.
   (d) For any positive real number y there is exactly one positive real number x such that $x^2 = y$.

§4. Ordered couples and Cartesian products

We write $\{a\}$ (read: the *singleton* of a) for the set whose only member is a, i.e., $\{x: x = a\}$. Likewise, $\{a, b\} = \{x: x = a \text{ or } x = b\}$, and $\{a, b, c\}$, etc., are understood similarly. Clearly the sets $\emptyset$ and $\{\emptyset\}$ are different, as one has a member, the other not. We see easily that all of the following are different: $\emptyset$, $\{\emptyset\}$, $\{\{\emptyset\}\}$, $\{\emptyset, \{\emptyset\}\}$, and $\{\{\{\emptyset\}\}\}$; and we could go on and on.

If $a \neq b$ then, nevertheless, $\{a, b\} = \{b, a\}$. Since at least the time of Descartes and coordinates in geometry, an 'ordered couple' (a, b) has been needed such that $(a, b) \neq (b, a)$ when $a \neq b$ and indeed, in general: if $(a, b) = (c, d)$ then $a = c$ and $b = d$. Although $\{a, b\}$ cannot serve for (a, b), the following 'clever' combination of a and b can:

4.1. Definition (a). $(a, b) = \{\{a\}, \{a, b\}\}$.
Theorem (b). If $(a, b) = (c, d)$ then $a = c$ and $b = d$.

The proof only requires a patient look at the possibilities; it takes perhaps half a page or so (see Problem 1).

The particular definition (a) of (a, b) is due to C. Kuratowski. N. Wiener had a few years earlier found another, which was a bit more complicated. The point here is that because of 4.1 we are still working with only the primitive notion '$x \in y$'; we did not have to add '(a, b)' as a second undefined notion (for which (b) would be an 'axiom'). It is interesting that such a famous mathematician as Wiener (or Kuratowski) was involved in the quite easy step of finding a definition like (a) which works. Of course, as often in mathematics, the hard step was to realize one should try to find such a definition!

The ordered couple is a simple but quite typical example of the way in which one incorporates in set theory (whose only primitive notion is $\in$) notions from the 'old' or ordinary mathematics, indeed, ultimately, all such notions. *There is no claim that the ordered couple* $(x, y) = \{\{x\}, \{x, y\}\}$ is the *very one* you had in mind before this, say, when doing analytic geometry. Rather the point is that the old ordered couple was not fully specified; and indeed the only thing specified was that the rule (or axiom) in 4.1 (b) holds. By Theorem 4.1(b), *our* ordered couple has this property, and that is the end of the matter. We shall soon (and easily) make exactly analogous replacements in set theory for several very important old notions, such as 'relation' and 'function' and the correlated notions "relation R *holds* between x and y", and "the value of f at x." Of course, 'function' is as central in mathematics as any other notion, the only competitor being 'set' itself!

The set $A \times B = \{(a, b): a \in A \text{ and } b \in B\}$ is called the *Cartesian product* of the (arbitrary) sets A and B. If A and B are intervals (say $[a_1, a_2]$ and $[b_1, b_2]$ in the reals then in (Descartes') analytic geometry, $A \times B$ is a rectangle in the plane. But the definition clearly makes sense for *any* sets A and B. There are a number of very easily proved laws regarding $\times$. Indeed almost anything seems to be true, as in the 'odd' law (d) below (cf. Halmos [1960]).

Theorem 4.2.
(a) $A \times \bigcup_{i \in I} \mathcal{B}_i = \bigcup_{i \in I}(A \times \mathcal{B}_i)$ *(and a 'dual' here and below with multiplication by 'A' on the right side).*

(b) $A \times \bigcap_{i \in I} \mathcal{B}_i = \bigcap_{i \in I}(A \times \mathcal{B}_i)$.
(c) $A \times (B - C) = A \times B - A \times C$
(d) $(A \cap B) \times (C \cap D) = (A \times C) \cap \ (B \times D)$.

Problems

1. Prove 4.1 (b).
2. Prove 4.2 (a), 4.2 (c), and infer 4.2 (b).
3. Prove 4.2 (d).

§ 5. Relations and functions

We say R is a *relation* if R is a set of ordered couples. ('R', 'S', and 'T' will denote relations.) We write Rxy (or perhaps xRy), and say R *holds between* x *and* y, if $(x, y) \in R$. Intuitively, a relation like fatherhood is something that holds or not for any given x, y. It is immediately clear that our notions will do just as well!

The *domain* of a relation R, Dom R, is defined to be $\{x\text{: for some } y, xRy\}$ and the *range*, Rng R, is $\{y\text{: for some } x, xRy\}$.

We say that f is a *function* if f is a relation and for any x, y, z if $(x, y), (x, z) \in f$ then $y = z$. For each $x \in \text{Dom} f$, we put $f(x)$ (read 'f *of* x') = the unique y such that $(x, y) \in f$. As usual, 'f', 'g', 'h' always denote functions. There are several alternate ways of saying things here. Thus 'f is *on* A' means Dom $f = A$; 'f is *to*, or *into* B means Rng $f \subseteq B$; 'f is *onto* B' means Rng $f = B$. Also, '$f: A \rightarrow B$' means f is a function on A to B. We have almost at once (taking $f = ((x, \mathcal{A}x)\text{: } x \in A)$):

Theorem 5.1. *There exists exactly one function f on A such that for each $x \in A$, $f(x) = \mathcal{A}_x$.*

(Of course 5.1 implies the same result with any naming expression replacing '$\mathcal{A}_x$'.) Obviously 5.1 is exactly what is needed to insure that our new notion of 'function' (and 'f of x') can perfectly replace the old-style 'function' which was given by 'specifying the domain set and a rule'.

It is convenient to have, in analogy with the set-builder notation '$\{-:---\}$', a function-builder notation. To date there has not been any general agreement on what notation to use. For the unique function f on X such that, for all $x \in X$, $f(x) = \mathcal{A}_x$, all of the following are in use:

$$(\mathcal{A}_x\text{: } x \in X)$$

$$(\lambda x \in X)\mathcal{A}_x$$

$$x \in X \mapsto \mathcal{A}_x.$$

If X is 'fixed', then one often deletes '$\in X$' in such notations. Incidentally, if we are thinking of the $\mathcal{A}_x$ as *sets* (officially, everything is a set, but in some discussions some sets 'act like atoms') then the *function* $f = (\mathfrak{A}_x : x \in X)$ is sometimes called an *indexed family* (of sets). The (plain) *family* $\{\mathfrak{A}_x : x \in X\}$ has 'less information.'

A function on the set of real numbers to itself can be pictured in a familiar, *precise* way. An arbitrary function, say on A to B, is often pictured in an unofficial and imprecise, but helpful, way, more or less as follows:

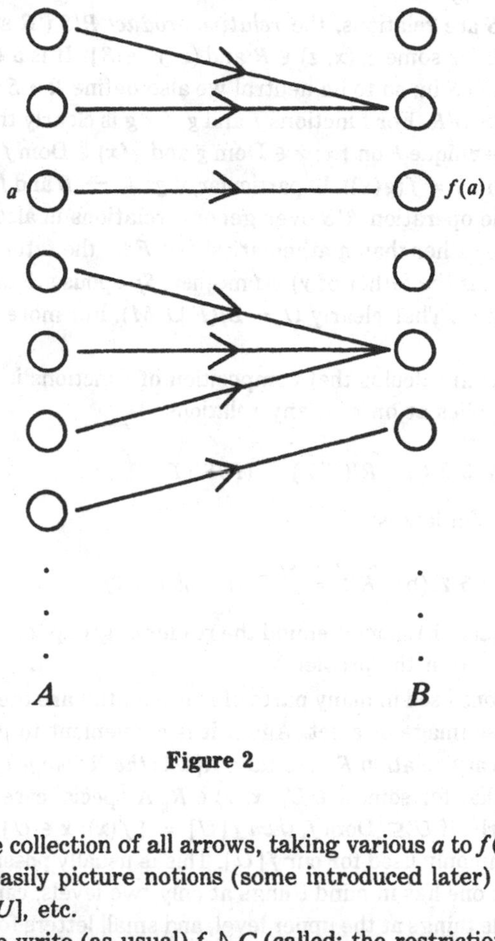

Figure 2

We think of f as the collection of all arrows, taking various a to $f(a)$. By such drawings one can easily picture notions (some introduced later) like one-to-one, onto, $f \circ g$, $f[U]$, etc.

If $C \subseteq \text{Dom } f$, we write (as usual) $f \restriction C$ (called: the restriction of f *to* C) for the function g on C such that, for any $x \in C$, $g(x) = f(x)$. (The form used in the definition illustrates the fact that (because of 5.1) we can use the old style in discussing functions when we wish – forgetting all about ordered couples, etc.).

Here is another, typical example of 'absorbing' old notions. By a *function of two variables* we mean a function f whose domain is a set of ordered couples. We agree to write $f(x, y)$ for $f((x, y))$!

Returning to general relations, we take $\breve{R}$ (*R converse*) to be the set of all (y, x) such that $(x, y) \in R$. Of course, a function f is said to be *one-to-one* if for any $x, y \in \text{Dom } f$, if $f(x) = f(y)$ then $x = y$. The converse of a function which is not one-to-one is a relation but not a function; this is one reason it is convenient to consider arbitrary relations. Clearly, f is one-to-one if and only $\breve{f}$ is a function. In this case, the *function* $\breve{f}$ is called the *inverse* function (of f) and written f^{-1}.

If R and S are relations, the *relative product* R/S ('R stroke S') is defined to be $\{(x, y)\text{: for some } z, (x, z) \in R \text{ and } (z, y) \in S\}$. It is a toss-up which order to write R and S in; so to be neutral we also define $R \circ S$ (the *composition* of R and S) to be S/R. For functions f and g, $f \circ g$ is clearly the familiar composition, i.e., the unique h on $\{x\colon x \in \text{Dom } g \text{ and } g(x) \in \text{Dom } f\}$ such that, for any $x \in \text{Dom } h$, $h(x) = f(g(x))$. In particular, if $g\colon A \to B$ and $f\colon B \to C$ then $f \circ g\colon A \to C$. The operation R/S over general relations is also familiar, but from ordinary life rather than mathematics! Let F be the father relation, (i.e., xFy means that x is the father of y), M mother, Sp spouse, B brother, S sister, and U (blood) uncle. Then clearly $U = B/(F \cup M)$. For more examples, see problem 1.

One learns in calculus that composition of functions is associative; in fact, the same applies at once to any relations:

Proposition 5.2.(a) $R/(S/T) = (R/S)/T$.

Another useful law is:

Proposition 5.2.(b) $\breve{\overline{R/S}} = \breve{S}/\breve{R}$ *(Problem 2).*

(Both (a) and (b) should remind the reader of group theory.) Some other laws involving / are in the problems.

Two notions used in many parts of mathematics are the f-image of a set and the f-inverse image of a set. Again it is convenient to discuss first the case of an arbitrary relation R. We take $R[U]$ (*the R-image of U*) to be the set of all y such that for some $x \in U$, $(x, y) \in R$. A special case is the f-image of U, $f[U]$. Clearly, if $U \subseteq \text{Dom } f$, *then* $f[U] = \{f(x)\colon x \in U\}$. (The notation $f(U)$ is very commonly used for our $f[U]$. This is usually possible, because in most discussions one has in mind things at only two levels; capital letters are used (say) for the things at the upper level, and small letters for things at the lower level. In this way the ambiguity between reading $f(U)$ as the *f-image* of U or as f *of* U is resolved. But for us this is not possible, as we consider sets at all levels.) Given a function f (not necessarily one-to-one), we can form the *relation* $\breve{f}$ and the image set $\breve{f}[V]$; this set is called the *f-inverse image* of V, and is written $f^{-1}V$. Clearly, $f^{-1}V = \{x \in \text{Dom } f\colon f(x) \in V\}$.

It is very important (for example, in topology, measure theory, etc.) that the 'direct' image goes through $\bigcup$ (and not much more), but (as is sometimes said), the inverse image 'goes through everything'! Thus:

Proposition 5.3. *Suppose for all $i \in I$, $\mathcal{A}_i \subseteq X$. Then*

(a) $R[\bigcup_{i\in I}\mathcal{A}_i] = \bigcup_{i\in I} R[\mathcal{A}_i]$ *(this includes both the f-image and the f-inverse image).*

(b) $f^{-1}(A - B) = f^{-1}A - f^{-1}B$.

(c) $f^{-1}(\bigcap_{i\in I}\mathcal{A}_i) = \bigcap_{i\in I} f^{-1}\mathcal{A}_i$, *if* $I \neq \emptyset$.

(c) follows from (a) and (b) at once using de Morgan's laws; the straightforward proofs of (a) and (b) are problems 3 and 4.

On any set A, the *identity function for A, Id_A,* is the function $x \in A \mapsto x$. Clearly, if $f: A \to B$ then $Id_B \circ f = f$ and $f \circ Id_A = f$. Let us agree to write $A \underset{f}{\sim} B$ to mean that f is a one-to-one function on A onto B; sometimes one says instead: *f is a one-to-one correspondence between A and B.* One sees in his head that

Proposition 5.4. (a) *(i)* $A \underset{Id_A}{\sim} A$; *(ii) if* $A \underset{f}{\sim} B$ *then* $B \underset{f^{-1}}{\sim} A$; *(iii) if* $A \underset{f}{\sim} B$ *and* $B \underset{g}{\sim} C$ *then* $A \underset{g\circ f}{\sim} C$.

Following Cantor we say that A and B are *(set-theoretically) equivalent* (or 'equinumerous'), written $A \sim B$, if, for some f, $A \underset{f}{\sim} B$ (or, in other words, if A and B can be put into one-to-one correspondence). Set-theoretical equivalence is the central notion of Cantor's set theory. From (a) we have at once

Proposition 5.4.(b) *(i)* $A \sim A$; *(ii) if* $A \sim B$ *then* $B \sim A$; *and (iii) if* $A \sim B$ *and* $B \sim C$ *then* $A \sim C$.

As is already familiar, a notion like $\sim$ for which 5.4(i) – (iii) hold is called an 'equivalence.'

Remark. Another notion, slightly different from 'function' as above, is sometimes more suggestive and often used; we shall call it 'mapping.' (The actual choice of words here is in a state of flux, but no matter.) We call F a *mapping* if it is of the form (f, B) where $\text{Rng} f \subseteq B$. In a new sense we now speak of Dom F to mean (the old) Dom f, and $F(x)$ to mean (the old) $f(x)$, and we write $F: A \to B$ to mean: for some f, $F = (f, B)$ and $f: A \to B$. F determines the two sets Rng f and B, which are variously called the exact range and the range of F or the range and the counterdomain of F. If $F: A \to B$ and $G: B' \to C$ we now only consider $G \circ F$ to be *defined* if $B = B'$, and then it is $(g \circ f, C)$. There is no difficulty in passing back and forth between the two groups of notions, once one is clear they are different. (We shall not use the 'mapping notation' again.)

Problems

1. Using some of F, M, Sp, B, and S (given before 5.2) and $\cup$, $\cap$, $\sim$, /, $\breve{}$, express G (grandmother), B and B' (brother-in-law in the narrow or wide sense, respectively).

2. Prove 5.2(b).
3. Prove 5.3(a).
4. Prove 5.3(b).

The results in the remaining problems 4 – 9 are not used directly in the later development. On the other hand, they happen to be a very instructive group of problems for students who are fairly new at finding and writing mathematical proofs.

5. Prove the following (a strong 'converse' for 5.3(b) and (c)): If, for any A, B, $R[A \cap B] = R[A] \cap R[B]$, then $\breve{R}$ is a function.

Here are some more laws to remember and prove:

6. $R/\bigcup_{i \in I} S_i = \bigcup_{i \in I} R/S_i$.
7. If R is a function then $R/\bigcap_{i \in I} S_i = \bigcap_{i \in I} R/S_i$.
8. $(R/S)[A] = S[R[A]]$.
9. Since the laws for R-image and $R/\_$ seem to be parallel, try to reduce the first to the second. Specifically, fill in the blank and prove: $\{z\} \times R[A] = \_/R$. (Here z is any fixed thing. (*Aside:* U and $\{z\} \times U$ are practically the same thing, so this does 'reduce' image to /.)

§ 6. Sets of sets, power set, arbitrary Cartesian product

We denote by B^C the set of all functions on C to B. (Some authors write ${}^C B$.) As will be seen in Chapter 2, if C has n elements and B has m, then B^C has m^n; hence the notation: B^C. (In forming B^C we are for the first time deliberately forming a set at a 'higher level'. Imagine that B and C are both infinite sets of apples. Then B^C is *not* a set of ordered couples of apples. Rather, each member of B^C is already an (infinite) set of ordered couples of apples and B^C itself lies at a higher level.)

A more general notion is the (arbitrary) *Cartesian product* $\prod_{i \in I} \mathcal{A}_i$, which is defined to be *the set of all f on I* such that, for each $i \in I$, $f(i) \in \mathcal{A}_i$. If $\mathcal{A}_i = B$ for all $i \in I$, then clearly $\prod_{i \in I} \mathcal{A}_i = \prod_{i \in I} B = B^I$. $\prod_{i \in I} \mathcal{A}_i$ also generalizes the old Cartesian product $A \times B$, although not quite directly. Indeed, let u and v be any two distinct things; let $\mathcal{A}_u = A$ and $\mathcal{A}_v = B$; and let $I = \{u, v\}$. The $\prod_{i \in I} \mathcal{A}_i$ is not identical to $A \times B$, but the two sets are a natural one-to-one correspondence – which takes $\{(u, b), (v, c)\}$ to (b, c) (if $b \in B$ and $c \in C$).

A third related notion is the set $P(A)$ of all subsets of the set A. It is also called the *power set of A* (hence the 'P'). Here 'power' means 'exponent'; indeed, taking again any distinct u and v, there is a *natural one-to-one* correspondence H between $P(A)$ and $\{u, v\}^A$. To define H, consider any $B \subseteq A$. In a familiar way, we form the 'characteristic function' c_B of B as follows: c_B is the function on A such that, for any $x \in A$, $c_B(x) = u$ if $x \in B$, and $c_B(x) = v$ if $x \notin B$. Then define $H(B) = c_B$. The reader will easily verify (see Problem 1) that

Proposition 6.1. $P(A) \,\widetilde{\overline{{}_H}}\, \{u, v\}^A$.

Using the Cartesian product $\Pi_{i \in I} A_i$ we can now state a distributive law much more general than 3.1(b) for $\bigcap$ and $\bigcup$.

Theorem 6.2^{AC}. *(Consider sets* $\mathcal{A}_{ij} \subseteq X$ *and take any* $\bigcap_{i \in \emptyset} B_i = X$.*)*

$$\bigcap_{i \in I} \bigcup_{j \in J_i} \mathcal{A}_{ij} = \bigcup_{f \in \prod_{i \in I} J_i} \bigcap_{i \in I} \mathcal{A}_{if(i)}.$$

(The dual follows by applying de Morgan's law.)

Proof. The proof of '$\supseteq$' is straightforward (see Problem 2). We prove '$\subseteq$'. Suppose x belongs to the 'left side'. Then for any $i \in I$ there exists j such that $j \in J_i$ and $x \in \mathcal{A}_{ij}$. Since for each $i \in I$, there is such a j, it obviously follows that there is a function f on I such that for each $i \in I$, $f(i)$ is such a j, i.e., $f(i) \in J_i$ and $x \in \mathcal{A}_{if(i)}$. Hence x belongs to the 'right side', since clearly for this very f, $f \in \Pi_{i \in I} J_i$ and, for all $i \in I$, $x \in \mathcal{A}_{if(i)}$. Thus '$\subseteq$' is proved.

In the course of this proof we have employed without mention, as obvious, the following principle, called the Axiom of Choice:

The Axiom of Choice (abbreviated AC). *Let F be a function on I such that for each $i \in I$, $F(i)$ is a non-empty set. Then there is a function f on I such that for each $i \in I$, $f(i) \in F_i$. (Note: Such an f is called a* choice function *(or* selector) *for the indexed family $(F(i) : i \in I)$.)*

At this point we do not *adopt AC*. We will indicate dependence on it as in '6.2^{AC}' above. Notice that the Axiom of Choice is identical to the statement: If for each $i \in I$, $F(i) \neq \emptyset$, then $\Pi_{i \in I} F(i) \neq \emptyset$.

The argument in the proof of 6.2 ($\subseteq$) above would be justified directly if we used the stronger *Axiom Schema of Choice* (AC^+) stated in Chapter 4. Using only AC, we must pause in the proof just before 'Since', and introduce a function F on I such that for each $i \in I$, $F(i) = \{j \in J_i : x \in \mathcal{A}_{ij}\}$. Then the next step *is* justified by AC. It happens here that for each i, the collection of j's to be 'chosen from' is small enough to be a set (indeed a subset of J_i). This is required in AC (but not in AC^+). *In practice*, most mathematicians think and write as in the original proof of '$\subseteq$' above (a standard justification using AC or AC^+ being 'always available'.) Indeed they would probably write: "Let f on I be a function which for each $i \in I$ *chooses* $F(i) \in J_i$ such that $x \in \mathcal{A}_{if(i)}$." Note however, that even in our proof of '$\subseteq$', and in AC itself (and all its applications) it is never *necessary* to use the words "choose" or "choice" at all; they are always used for heuristic purposes only.

If we have a single set Q known to have a member, then ordinary logic allows us to reason in a way that amounts to 'choosing' an $x \in Q$. Indeed we shall prove in Chapter 2 that AC holds whenever the set I is finite. Thus the point of the axiom of choice is when we are asked to 'make infinitely many choices.' Even when I is infinite, AC is often not needed. Two famous contrasting examples were given by Bertrand Russell. Let I be infinite. First, suppose each $F(i)$ is a (different) pair of shoes. Then without AC we can take (for each $i \in I$) $f(i)$ to be the left shoe of the pair $F(i)$. Second, suppose each $F(i)$ is a pair of socks. Now AC is apparently the only way to get a choice function f! Incidentally, notice that having the sets $F(i)$ all finite does not seem to help.

The Axiom of Choice was first stated by Zermelo [F1904]. (Earlier, it had been used implicitly by a number of people, including Cantor.) Many, perhaps most, mathematicians find the Axiom of Choice to be clearly 'true.' Nevertheless, the years since 1904, some very great mathematicians have argued, often bitterly, that the Axiom of Choice is dubious or outright 'false' – for example, Poincaré, Borel, Brouwer, and others. (A full account of the history of the Axiom of Choice can be found in the very readable new book of Gregory Moore [1982], *Zermelo's Axiom of Choice – Its Origin, Development, and Influence.*)

As we mentioned in discussing Extensionality, there are various ways in which one might interpret 'set'. Roughly speaking, the Axiom of Choice seems to be true if one is allowing (as we are) a set A say of real numbers to be those hit by a bolt of lightning. But if instead we demand that some 'rule' be used in determining the members of A, then, as in the socks example, the axiom of choice might well be false.

A great deal of modern mathematics depends on the axiom of choice. Our aim in this book will be mainly to discuss what can be proved using the Axiom of Choice. However, the situation is not so simple. Consider, e.g., the Cantor-Bernstein Theorem: *If* $F: A \to B$ *and* $G: B \to A$ *and both are* one-to-one, *then there is* a function H *which is* one-to-one *on* A *onto* B. Cantor proved it using (intuitively) the choice axiom. Bernstein gave a beautiful proof (see Chapter 4) without the choice axiom. Cantor's argument only shows very indirectly that an H must *exist*; whereas Bernstein's proof shows how to construct an H from F and G. Thus Bernstein's proof avoiding AC is superior, in a way, even if we are assuming AC. This situation (hard to characterize precisely) is very common. Consequently, *when it is easy to do so*, we shall at least for a while try to avoid AC.

There are other infinite distributive laws, for example (see Problem 3):

Proposition 6.3.^{AC} *(Assume all* $\mathcal{A}_{ij} \subseteq x$ *and take* $\bigcap_{\iota \in \emptyset} \mathcal{B}_\iota = X$*)*

$$\prod_{i \in I} \bigcup_{j \in J_i} \mathcal{A}_{ij} = \bigcup_{f \in \prod_{i \in I} J_i} \prod_{i \in I} \mathcal{A}_{if(i)}$$

Problems

1. Prove 6.1.
2. Prove 6.2 ($\supseteq$).

3. Prove 6.3.
4. Does 6.3 hold if $\bigcup$ is replaced everywhere by $\bigcap$?
5^{AC}. Let f be on A onto B. Show there exists $g\colon B \to A$ such that for each $b \in B, f(g(b)) = b$. (A corresponding *mapping* is called a right-inverse). Show also that g is one-to-one.

§ 7. Structures

We say that $\circ$ is a *binary operation* over A if $\circ\colon A \times A \to A$. (As usual, $\circ((x, y))$ is written $x \circ y$.) R is called a *binary relation over* A if $R \subseteq A \times A$. A typical (algebraic) structure is something of the form $\underline{A} = (A, R, \circ)$ where $\circ$ and R are over A). The set A is called the *universe* of $\underline{A}$. A *structure* can have any number of R's and $\circ$'s and these can have any number of places (not just two), and there can also be distinguished elements. Thus an ordered field is of the form $(A, +, \cdot, <, 0, 1)$. Except in the (informal) Chapter 3, our interest will henceforth be with *structures* (A, R), R binary. (So we have not even discussed formally the notion of say a 3-place relation or operation; see §3 of Chapter 2.) Incidentally, it is convenient for us to allow structures (A, R) whose universe A is empty.

Let $\underline{A} = (A, R)$ and $\underline{A}' = (A', R')$ (assumed, as always, to be structures, i.e., $R \subseteq A \times A$ and $R' \subseteq A' \times A'$). We say that f is an $(\underline{A}, \underline{A}')$-*isomorphism* (or an isomorphism between $\underline{A}$ and $\underline{A}'$), in symbols, $\underline{A} \underset{f}{\equiv} \underline{A}'$, provided that $A \;_{\overline{f}}\, A'$ and for any $x, y \in A$, $x R y$ if and only if $f(x) R' f(y)$. (If instead $\underline{A} = (A, \circ')$, $\underline{A}' = (A', \circ')$, one has instead the familiar condition: $f(x \circ y) = f(x) \circ' f(y)$.) We write $\underline{A} \cong \underline{A}'$ and say that $\underline{A}$ is isomorphic to $\underline{A}'$ if for some f; $\underline{A} \underset{f}{\equiv} \underline{A}'$. (The notion of isomorphism extends to arbitrary algebraic structures and even to non-algebraic structures like topological spaces, manifolds, etc.)

We can picture a one-to-one correspondence f between A and A' as follows:

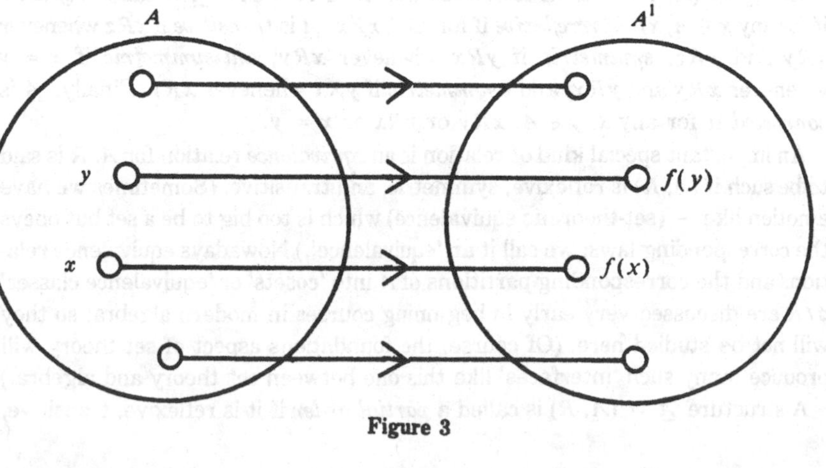

Figure 3

For f to be an $((A, R), (A', R'))$-isomorphism, the 'behavior' in general of $f(x)$, $f(y)$, etc. in (A', R') must be identical with that of x, y, etc. in (A, R). Thus instead of '$\underline{A}$ and $\underline{A}'$ are isomorphic', one sometimes says that $\underline{A}$ and $\underline{A}'$ are 'abstractly identical'.

Proposition 7.1. *The laws 5.4 (a) and (b) about '$A \underset{f}{\sim} B$' and '$A \sim B$' extend at once to '$\underline{A} \underset{f}{\cong} \underline{B}$' and '$\underline{A} \cong \underline{B}$'.*

Cantor's '$A \sim B$' is, of course, the special case of isomorphism between structures having 'no structure', i.e., no R's or $\circ$'s.

Given a structure $\underline{A} = (A, R)$, the idea of a substructure is even easier than the familiar one for groups, rings, etc. Indeed given *any* subset B of A (which need not be 'closed' in some sense), the structure $(B, R \cap (B \times B))$ is determined. ($R \cap (B \times B)$ is just R, 'considered only over B'.) This structure is written $\underline{A} \mid B$, and structures like this for various $B \subseteq A$ are the *substructures* of $\underline{A}$.

We say that f is an isomorphism of $\underline{A} = (A, R)$ *into* $\underline{A}' = (\underline{A}', \underline{R}')$ (a new notion) if f is one-to-one on A *into* A' and for any $x, y \in A$, xRy if and only if $f(x) R' f(y)$. A simple but useful fact is:

Proposition 7.2. *f is an isomorphism of $\underline{A}$ into $\underline{A}'$ if and only if f is an isomorphism between $\underline{A}$ and some substructure of $\underline{A}'$.*

Problems

1. Write out the six laws stated in 7.1.
2. Prove 7.2.
3. Prove that if $\underline{A} = (A, R)$ and $A \underset{f}{\sim} A'$ (A' is just a set) then there is exactly one R' such that $(A, R) \underset{f}{\cong} (A', R')$.

§ 8. Partial orders and orders

Let $\underline{A} = (A, R)$ be a structure (so that $R \subseteq A \times A$). $\underline{A}$ is called *reflexive* if for any $x \in A$, xRx; *irreflexive* if for all x, $x \not R x$. $\underline{A}$ is *transitive* if xRz whenever xRy and yRz; *symmetric* if yRx whenever xRy; *antisymmetric* if $x = y$ whenever xRy and yRx; and *asymmetric* if $y \not R x$ whenever xRy. Finally, $\underline{A}$ is *connected* if for any $x, y \in A$, xRy or yRx or $x = y$.

An important special kind of relation is an *equivalence* relation for A. R is said to be such if (A, R) is reflexive, symmetric, and transitive. (Sometimes we have a notion like $\sim$ (set-theoretic equivalence) which is too big to be a set but obeys the corresponding laws; we call it an 'equivalence'.) Nowadays equivalence relations and the corresponding partitions of A into 'cosets' or 'equivalence classes' a/R are discussed very early in beginning courses in modern algebra; so they will not be studied here. (Of course, the foundations aspect of set theory will produce many such 'interfaces' like this one between set theory and algebra.)

A structure $\underline{A} = (A, R)$ is called a *partial order* if it is reflexive, transitive, and antisymmetric. $\underline{A}$ is called a *strict partial order* (a variant) if $\underline{A}$ is irreflexive and transitive. $\underline{A}$ is called an *order* (a stronger notion) if it is a partial order and is connected. As a variant, $\underline{A}$ is called a *strict order* if it is a strict partial order and is connected.

In a familiar way, we often use '$\leq$' as a variable, instead of 'R', when (A, R) is reflexive, and '$<$' when (A, R) is irreflexive. Moreover, if we start with $(A, <)$, then we write $x \leq y$ to mean: $x < y$ or $x = y$; and if we start with $(A, \leq)$, then we write $x < y$ to mean $x \leq y$ and $x \neq y$. (If one does first one then the other he clearly gets back where he started.) It is very easy to check the following: If $\underline{A} = (A, \leq)$ is a partial order (or order) then $(A, <)$ is a strict partial order (resp., strict order). On the other hand, if $(A, <)$ is a strict partial order (or strict order) then $(A, \leq)$ is a partial order (resp., order). (Having been so careful we shall henceforth be sloppy in the usual ways. Just for an example, we might say "let $(A, <)$ be an order" and expect the reader to understand 'strict order', because we wrote '$<$'.)

Thus we are interested in two kinds of structures, orders and (the more general) partial orders. Here are some unofficial examples (which use material outside of our present development): If $\leq$ is in each case the usual one, then $(N, \leq)$, $(Z, \leq)$, $(Q, \leq)$, $(R, \leq)$, are orders. Here, as usual, $N(Z, Q$, or $R)$ is the set of all natural numbers (integers, rationals, or reals, respectively). If $m, n \in N$ let $m \mid n$ mean m divides n, (i.e., for some $r \in N$, $mr = n$). Then $(N, |)$ is a partial order (not an order).

Within our own development, here is one important example: For any family Q of sets, consider the inclusion relation cut down to Q, i.e., the relation $\subseteq_Q = \{(A, B): A, B \in Q \text{ and } A \subseteq B\}$. Obviously:

Proposition 8.1. *$(Q, \subseteq_Q)$ is a partial order.* (For a 'converse' see Problem 1.)

In this way one can picture many examples of partial orders, for instance let Q have as members just the three sets shown below:

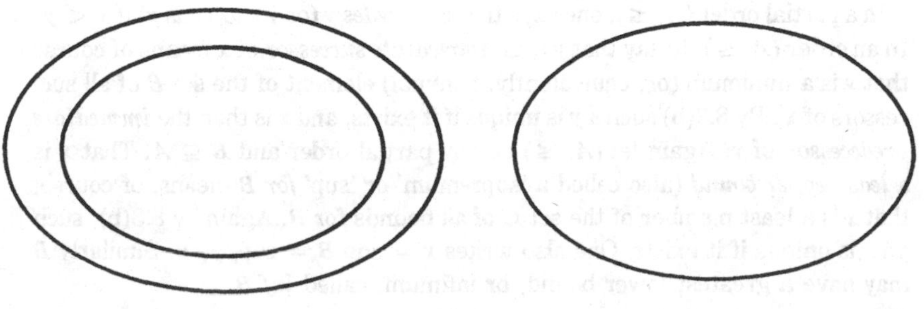

Figure 4

Obviously this gives a (three element) partial order which is not an order.

A family Q of sets which *is* ordered by $\subseteq_Q$ is called a *chain.*

Orders sometimes behave in a special way (among general structures). For example, as one easily sees:

Proposition 8.2. *Suppose (A, R) and (A', R') are strict orders, f is on A onto A' and 'order preserving', i.e., $f(x)R'f(y)$ whenever xRy. Then $(A, R) \underset{f}{\cong} (A', R')$. (For the proof and an improvement see Problem 2.)*

Obviously, if $\underline{B}$ is a substructure of $\underline{A}$ and $\underline{A}$ is a partial order, strict partial order, order, or strict order, then so is $\underline{B}$. The same is true if one 'turns (A, R) around,' that is: If (A, R) is a partial order, etc., as above, then so is $(A, \breve{R})$. (Of course, as usual, we write $\geq$ for $\breve{\leq}$.)

Let $\underline{A} = (A, R)$ be at least either reflexive or irreflexive, so we can in an obvious way understand $\leq^{\underline{A}}$ and $<^{\underline{A}}$ (written often as just $\leq$ and $<$). A simple, but often blurred distinction is made in the following definitions: If $B \subseteq A$ and $x \in A$, we say that x is a *minimal element* of B ('over $\underline{A}$' is understood) if $x \in B$ and for all $y \in B$, $y \nless x$. On the other hand, we say x is a *minimum element* of B if $x \in B$ and for all $y \in B$, $x \leq y$. Also, x is called a *lower bound* for B if for all $y \in B$, $x \leq y$. The notions *maximal, maximum, upper bound* are obtained by passing from $\leq$ to $\geq$; and everything about 'minimal', etc., below, applies analogously to 'maximal', etc.

There are an incredible number of synonyms for 'minimum' in common usage – for example, least, smallest, first, leftmost, ...! The following propositions have easy proofs (Problem 3), but are well worth noting:

Proposition 8.3. *Let $\underline{A} = (A, \leq)$ be a partial order, and $B \subseteq A$.*

(a) *'Minimum' implies 'minimal' (i.e., if x is a minimum element of B then x is a minimal element of B).*

(b) *$\underline{B}$ has at most one minimum element.*

(c) *If $\underline{A}$ is an order, then 'minimum' = 'minimal' (i.e., x is a minimum element of B if and only if x is a minimal element of B).*

In a partial order $(A, \leq)$, one says that x *precedes* y (or y succeeds x) if $x < y$. In an order $(A, \leq)$, to say that y is an *immediate successor* of x means of course that y is a minimum (or, equivalently, minimal) element of the set B of all successors of x. By 8.3(b) such a y is unique if it exists, and x is then the *immediate predecessor* of y. Again let $(A, \leq)$ be any partial order and $B \subseteq A$. That x is a *least upper bound* (also called a 'supremum' or 'sup' *for* B means, of course, that x is a least member of the set C of all bounds for B. Again by 8.3(b), such an x is unique if it exists. One also writes $x = \sup B = \sup_{u \in B} u$. Similarly B may have a greatest lower bound, or infimum, called *inf* B.

Let $\underline{A} = (A, \leq)$ be an order. By an *initial segment* of $\underline{A}$ we mean any subset

B of A such that for any x, y, if $y < x$ and $x \in B$ then $y \in B$. Each element a of A determines two initial segments of $\underline{A}$, namely, Pred a (in full, $\text{Pred}^{<}_{\underline{A}} a) = \{x: x < a\}$ and $\text{Pred}^{\leq} a = \{x: x \leq a\}$. *The substructure of* $\underline{A}$ *whose universe is* Pred a *will be called* $\underline{A}_a$. (There is a corresponding notion of final segment, etc.) An initial segment B is called *proper* if $B \neq A$. Note that in $(Q, \leq)$, the set $\{x : x < \pi\}$ is an initial segment not of either form Pred a or $\text{Pred}^{\leq} a$ for $a \in Q$.

Given $U, V \subseteq A$ we say that U and V are *cofinal* (or confinal or coterminal, or that U is cofinal with V or that V is with U), all in the sense of the order $\underline{A}$, if for any $x \in U$ there is a $y \in V$ such that $x \leq y$, and vice versa (i.e., if $x \in V$ then for some $y \in U$, $x \leq y$). Any $U \subseteq A$ determines the set $\{x$: for some $y \in U$, $x \leq y\}$, which is clearly the $\subseteq$-smallest initial segment of $\underline{A}$ cofinal with U.

It is natural to consider certain kinds of orders $\underline{A} = (A, \leq)$. $\underline{A}$ is called *dense* if it has at least two elements and for any x and y, if $x < y$, then for some z $x < z < y$. $\underline{A}$ is called *discrete* if every element not a last element has an immediate successor and every element not a first has an immediate predecessor. $\underline{A}$ is called a *well-order* if every non-empty subset of A has a first element. See Problem 8 for easy examples. The detailed study of well-orders is one of the deepest parts of Cantor's set theory (see Chapter 7). But that does not prevent us from giving now the easy definition, and making various easy remarks about well-orders whenever they come along. Thus, it is immediate that *any substructure* (or suborder, as is said) *of a well-order is a well-order.* Also:

Proposition 8.4. *Let* $\underline{A} = (A, \leq)$ *be a well-order.*

(a) *If A is not empty there is a first element.*

(b) *Every element not a last has an immediate successor.*

(c) *Every proper initial segment is of the form Pred a (for some a).*

(a) and (b) are obvious, (c) is easy (see Problems 4 and 5).

We could and, strictly speaking, probably should have discussed above 'partial orders' and 'orders' whose universe is not a set (e.g., $\subseteq$ over *all* sets). Most results (and proofs) above hold in that greater generality. For some reason, the notation (which we actually have) needed to do this is very little used and unfamiliar, so to do it seems to be more trouble than it is worth. But we may fudge later and use a result (like 'maximum implies maximal') in the general case.

Problems

1. Prove that every partial order $(A, \leq)$ is isomorphic to $(Q, \subseteq_Q)$ for some Q. (Such a result is called a representation theorem.) Hint: Take $Q = \{\{b: b \leq a\}: a \in A\}$, and define the obvious isomorphism F.
2. Prove 8.2. Then, looking at your proof, improve 8.2 by weakening its

hypotheses (to something *like* "$\underline{A}$ is transitive and $\underline{A}'$ is antisymmetric"); and prove the improvement.

3. Give proofs of 8.3(a), (b), (c).

4. Prove 8.4(c). (In Problems 4 and 5 make up the right non-empty set to take the first element of.)

5. Prove that for any order $\underline{A} = (A, \leq)$, the following are equivalent:

   (a) $\underline{A}$ is a well-order.

   (b) Every non-empty final segment of $\underline{A}$ has a least element.

   (c) Every proper initial segment of $\underline{A}$ is of the form Pred a (for some $a \in A$).

6. Let $\underline{A} = (A, \leq)$ be any order. Show the following are equivalent:

   (a) Every non-empty subset of A bounded above has a least upper bound.

   (b) Every non-empty subset of A bounded below has a greatest lower bound.

   (c) Given any non-empty initial segment U having a non-empty complement V, either U has a last element or V has a first. (Otherwise, the pair (U, V) was called a *gap* by Dedekind.)

   (An order having (a), (b), or (c), and also dense is called a *continuous order*).

7. Assume knowledge of the usual orders $\underline{R} = (R, \leq)$ of the real numbers and $\underline{Q} = (Q, \leq)$ of the rationals. Show true or false:

   (a) Every non-empty proper initial segment of $\underline{R}$ is Pred a, for some $a \in R$.

   (b) Every non-empty proper initial segment of $\underline{R}$ is $\text{Pred}^{\leq} a$, for some $a \in R$.

   (c) Every non-empty proper initial segment of R is Pred a or $\text{Pred}^{\leq} a$, for some $a \in R$.

   (d) Every non-empty proper initial segment of $\underline{Q}$ is Pred a or $\text{Pred}^{\leq} a$ (in $\underline{Q}$ for some $a \in Q$.

8. Which of $\underline{N}$, $\underline{Z}$, $\underline{Q}$, $\underline{R}$ are dense? discrete? well-ordered? continuous (see problem 6)?

2. CARDINAL NUMBERS AND FINITE SETS

In this chapter we begin the study of Cantor's notions of set-theoretical equivalence and cardinal numbers. This study is carried further in Chapters 4 and 7.

§ 1. Cardinal numbers, +, and ≤

We saw in 1.5.4(b) (which means 5.4(b) of Chapter 1) that $\sim$ (set-theoretic equivalence) is an equivalence notion. Given such a notion – for example, 'triangle T is *similar* to triangle T'', we create in ordinary speech a corresponding notion 'the *shape* of a triangle' of which we only know (or care) that, for any triangles T and T',

the shape of T = the shape of T'
if and only if T is similar to T'.

It therefore seems reasonable to imagine as we now do that to each set A corresponds something called the *cardinal number of* A (written $\overline{\overline{A}}$), the overall correspondence being such that for any sets A, B:

1.1 $$\overline{\overline{A}} = \overline{\overline{B}} \text{ if and only if } A \sim B.$$

Cardinals or cardinal numbers (i.e., things of the form $\overline{\overline{A}}$ for some A) are denoted by 'κ', 'λ', and 'μ'. It is common to say sometimes 'the *power* of A' instead of 'the cardinal number of A'. We can now define 0, 1, and 2 by putting $0 = \overline{\overline{\emptyset}}$; $1 = \overline{\overline{\{x\}}}$ (clearly the same for all x), and $2 = \overline{\overline{\{x, y\}}}$ if $x \neq y$ (again clearly justified). Obviously (why?) 0, 1, and 2 are distinct.

Following Cantor we now define certain notions among cardinal numbers. Usually one or two justifying lemmas are needed before the natural definition can be made.

1.2 (**Definition of** $+$).

Proposition (a) *If* $A \sim A'$, $B \sim B'$, $A)(B$, *and* $A')(B'$, *then* $A \cup B \sim A' \cup B'$.

Proposition (b) *For any* X, Y *there exist* X', Y', *disjoint, such that* $X \sim X'$ *and* $Y \sim Y'$.

Definition (c) $\kappa + \lambda =$ *the unique* μ *such that for some* A, B, $\overline{\overline{A}} = \kappa$, $\overline{\overline{B}} = \lambda$, $A)(B$, *and* $\overline{\overline{A \cup B}} = \mu$. *(A unique* μ *always exists, by (a) and (b).)*

Proof. In (a), let $A \underset{f}{\sim} A'$ and $B \underset{g}{\sim} B'$. The reader will easily verify that $A \cup B \underset{f \cup g}{\sim} A' \cup B'$, which implies (a). In (b) a trick is needed. Let u and v be any distinct things (e.g., $u = \emptyset$, $v = \{\emptyset\}$). Given X, Y put $X' = \{u\} \times X$ and $Y' = \{v\} \times Y$. Clearly $X \sim X'$, $Y \sim Y'$, and $X')(Y'$.

Some facts about $\sim$ are clear without proof, as they are analogous to the proposition: "if the groups $\underline{G}$ and $\underline{G}'$ are isomorphic and $\underline{G}$ is Abelian, so is $\underline{G}'$". For example: if $\overline{\overline{Z}} = \kappa + \lambda$, then there are disjoint X, Y such that $Z = X \cup Y$, $\overline{\overline{X}} = \kappa$. and $\overline{\overline{Y}} = \lambda$.

Here are some readily proved facts about addition of cardinals.

Proposition 1.3.
(a) $\kappa + \lambda = \lambda + \kappa$.
(b) $\kappa + (\lambda + \mu) = (\kappa + \lambda) + \mu$.
(c) $\kappa + 0 = \kappa$.
(d) *If* $\kappa \neq 0$ *then for some* λ, $\kappa = \lambda + 1$.
(e) *If* $\kappa + 1 = \lambda + 1$ *then* $\kappa = \lambda$.

(a)-(d) can be seen at once 'in one's head'. In (e), let $\overline{\overline{A}} = \kappa + 1 = \lambda + 1$. Then there are $x, y \in A$ such that $\overline{\overline{A - \{x\}}} = \kappa$ and $\overline{\overline{A - \{y\}}} = \lambda$. If $x = y$, $\kappa = \lambda$ at once. If not, let $\mu = \overline{\overline{A - \{x, y\}}}$. Then clearly $\kappa = \mu + 1$ and $\lambda = \mu + 1$, so $\kappa = \lambda$.

The important notion '$\kappa \leq \lambda$' has many equivalent definitions. The one we now (first) give reduces $\leq$ algebraically to $+$.

Definition 1.4. *We say that* $\kappa \leq \lambda$ *if for some* μ, $\lambda = \kappa + \mu$.

Proposition 1.5. *The following are equivalent:*
(a) $\kappa \leq \lambda$.
(b) *For some* A, B, $\overline{\overline{A}} = \kappa$, $\overline{\overline{B}} = \lambda$, *and* $A \subseteq B$.
(c) *For any set* B *of power* λ *there exists* $A \subseteq B$ *of power* κ.
(d) *For any* A, B *of respective powers* κ, λ, *there exists* f *one-to-one on* A *to* B.
(e) *For any set* A *of power* κ *there exists* $B \supseteq A$ *of power* λ.

The reader can easily verify the equivalence of (a)-(d). (e) (a kind of dual of (c)) requires for its proof a result called the

Exchange Principle. For any X, Y there exists $Z)(X$ such that $Z \sim Y$.

This intuitively reasonable proposition is in fact a little mysterious, as it says that we can build things out in the unknown world! (Algebraists need the Exchange Principle to establish such theorems as: every field has an extension which is algebraically closed.) Problem (1) gives a hint how to prove the Exchange Principle. In problem 2, one sees that the equivalence of (e) with the others in 1.4 is routine using the Exchange Principle.

Some laws for $\leq$ are as follows (of course, $\kappa < \lambda$ means $\kappa \leq \lambda$ and $\kappa \neq \lambda$).

Proposition 1.6.
(a) $\kappa \leq \kappa$.
(b) *If* $\kappa \leq \lambda$ *and* $\lambda \leq \mu$ *then* $\kappa \leq \mu$.
(c) *If* $\kappa \leq \lambda$ *then* $\kappa + \mu \leq \lambda + \mu$.
(d) $0 \leq \kappa$.
(e) *If* $\kappa < \lambda$ *then* $\kappa + 1 \leq \lambda$.

The very easy proofs are left to the reader. Usually there are two proofs – one going back to $\sim$, the other making an algebraic reduction, using the definition of $\leq$, to laws about $+$.

1.6(a) and (b) give two of the four assertions needed to show $\leq$ 'orders' the cardinals. ('Orders' is in quotes because there are too many cardinals to form a set.) That the third condition: 'if $\kappa \leq \lambda$ and $\lambda \leq \kappa$ then $\kappa = \lambda$' holds is the well-known Cantor-Bernstein Theorem, which will be proved in § 1 of Chapter 4. The fourth condition '$\kappa \leq \lambda$ or $\lambda \leq \kappa$' (comparability of cardinals) can only be proved using the Axiom of Choice. It is one of Cantor's deepest results, and will be proved in Chapter 7.

Here is one last proposition about $\leq$:

Proposition 1.7. *For any* κ, $\{\lambda : \lambda \leq \kappa\}$ *exists.*

Proof. Let $\overline{\overline{A}} = \kappa$. The desired set is $\{\overline{\overline{B}} : B \in P(A)\}$, which exists by the Axiom of Replacement.

There is another natural way of saying that κ is small compared to λ. Tarski gave it the name $\kappa \leq^* \lambda$. Indeed, we write $\kappa \leq^* \lambda$ if, for some A, B, $\overline{\overline{A}} = \kappa$, $\overline{\overline{B}} = \lambda$, and *there is a function f on B onto A* (A is 'an image of B') or else $\kappa = \emptyset$. It is easy to see that

Proposition 1.8.
(a) *If* $\kappa \leq \lambda$ *then* $\kappa \leq^* \lambda$ *(see Problem 3).*

By Problem 5 of § 6 of Chapter 1, we see at once that the converse holds, assuming AC:

(b)^{AC} *If* $\kappa \leq^* \lambda$ *then* $\kappa \leq \lambda$.

An easy variant of (b) is also useful:

(c) *Suppose* $\overline{\overline{B}} = \lambda$ *and some R well-orders A. Then* $\kappa \leq^* \lambda$ *implies* $\kappa \leq \lambda$. *(See Problem 4.)*

Problems

1. Prove the Exchange Principle. Hint: take $t \notin \{x : \text{for some } y, (x, y) \in X\}$ and then use ordered couples.
2. Show that in 1.5, (e) is equivalent to (a).
3. Prove 1.8(a).
4. Prove 1.8(c). Hint: Avoid AC by choosing the R-least a such that whatever.

§ 2. Natural numbers and finite sets

We can now define the notion 'finite'. In a short time (as with, e.g., the notion 'function' in Chapter 1) we will in effect have back the use of all our old intuition about 'finite'!

Definition 2.1. κ *is a* natural number *if* κ *belongs to every set X such that* $0 \in X$ *and, for any* λ, *if* $\lambda \in X$ *then* $\lambda + 1 \in X$.

(As usual, 'j', 'k',..., 'n' will denote natural numbers.)

We say a set A is *finite* if $\overline{\overline{A}}$ is a natural number; A is *infinite* otherwise. (By 'abuse of language', one sometimes also calls κ a finite or infinite cardinal if it is the cardinal of a finite or infinite set.)

Our definition of 'natural number' is essentially one due to Bertrand Russell. We shall learn later some of the very many known equivalent definitions.

Theorem 2.2.

(a) 0 *is a natural number.*

(b) *If* κ *is a natural number so is* $\kappa + 1$.

(c) *(Induction) Suppose that* $\mathcal{P}\,0$ *('*$\mathcal{P}$ *holds for* 0'*); and that, for any* n, *if* $\mathcal{P}\,n$ *then* $\mathcal{P}(n + 1)$. *Then for every* n, $\mathcal{P}\,n$.

(Clearly (c) justifies the familiar method of 'proof by induction'.)

Proof. (a) and (b) are very easy (see Problem 1). For (c), suppose that the whole hypothesis of (c) holds, but that, for some particular n, $\mathcal{P}\,n$ fails. Put $X = \{m \leq n : \mathcal{P}\,m\}$. ($X$ exists since $X = \{\lambda : \lambda \leq n$ and λ is a natural number and $\mathcal{P}\,\lambda\}$, which exists by 1.7 and the Separation Axiom.) Obviously, $0 \in X$ (by (a) and 1.6(d)). It will be enough to show that: $\lambda + 1 \in X$ whenever $\lambda \in X$ – as then X is 'an X' as in Definition 2.1, so, by 2.1, the natural number $n \in X$, and so $\mathcal{P}\,n$ holds, a contradiction. Suppose then that $\lambda \in X$, so that $\lambda \leq n$, λ is a natural number, and $\mathcal{P}\,\lambda$. By our hypothesis (in (c)), $\mathcal{P}(\lambda + 1)$. By (b), $\lambda + 1$ is a natural number. Also, $\lambda < n$, as $\mathcal{P}\,n$ fails. Hence $\lambda + 1 \leq n$, by 1.6(e). So $\lambda + 1 \in X$, as desired.

If the set of all natural numbers exists, we call it N. But it is not necessary for us to assume now that N exists. The assumption that N exists is a form of what is called the Axiom of Infinity, which will be discussed in § 5 below.

Remark (which at this point must be taken rather loosely). In 2.1 and 2.2 (especially 2.2(c)) and a little later in discussing 'recursion' (4.1(b)), we use methods, both in formulation and proof, which do not require knowing that N (or any 'infinite' set) exists. Thus in the proof of 2.2(c), the set X we constructed was 'finite'. These methods are sometimes called the 'from below' approach. In the often used 'from above' approach one deals with natural numbers and finite sets by a 'detour or jump' through infinite sets, especially N. One has to go way out of his way to find a case where the 'above' method is superior to the 'below', but the reverse is often true. (Probably the popularity of the 'above' methods goes back to the fact that they were what was first discovered by Dedekind and Peano.)

Theorem 2.2 completely characterizes being a natural number, so that one can even forget Definition 2.1. This state of affairs is made precise and proved in Problem 2.

By Theorem 2.2(c), we can in the familiar way make proofs by induction. That is, if we want to prove that all natural numbers n have the property $\mathcal{Q}$, it is enough to prove $\mathcal{Q}0$ and for all n, if $\mathcal{Q}n$ then $\mathcal{Q}(n+1)$. We say at the beginning of such a proof: "We will now prove that for all n, $\mathcal{Q}n$, *by induction*" (or "*by induction on 'n'*").

Aside. Sometimes we say we will prove a certain theorem 'by induction', meaning only by (some application of) the method of induction. Even if that theorem has the form 'for all n, $\mathcal{P}n$' it may not be possible to prove this directly by induction. Rather you must conjecture another law 'for all n, $\mathcal{Q}n$', such that (1) it can be proved directly by induction, and (2) 'for all n, $\mathcal{P}n$' can be inferred from 'for all n, $\mathcal{Q}n$'. Of course at a certain point we will say in the exact way: "We will now prove for all n, $\mathcal{Q}n$, by induction on 'n' ".

We can now prove a rather large number of facts (mostly familiar) about natural numbers and finite sets.

Proposition 2.3.

(a) *$m + n$ is a natural number (the natural numbers are closed under $+$).*

Proof. We show that for all n, $m + n$ is a natural number, by induction on 'n'. (This example explains the use of 'on 'n'', which means here: with m fixed.) By 1.3(e), $m + 0 = m$, which is a natural number. Suppose $m + n$ is a natural number. Then by 1.3(b), $m + (n + 1) = (m + n) + 1$, which is a natural number by the inductive hypothesis and 2.2(b).

The rest of 2.3 and 2.4 are all proved (in the Problems) by induction except those called 'Corollary' which follow from earlier ones (with no use of induction).

(b) *If $\kappa \leq n$ then κ is a natural number. (Hint: You may need some propositions which could have been added to 1.6(a)-(e); but these are not hard to prove.) Of course, by (b), any subset of a finite set is finite.*

(c) *If $\kappa + n = \lambda + n$ then $\kappa = \lambda$. (Note that this includes but goes beyond the familiar case where κ is a natural number).*

(d) **Corollaries:**

  (*i*) *If $m \leq n$ then there is exactly one k such that $m + k = n$ (Subtraction.)*

  (*ii*) *A finite set is not equivalent to a proper subset of itself.*

  (*iii*) $n < n + 1$.

  (*iv*) *If $m \leq n$ and $n \leq m$ then $m = n$ (equivalently, if $m < n < p$ then $m < p$).*

(e) *$\kappa \leq n$ or $n \leq \kappa$. (So, using (b), if κ is infinite, $n < \kappa$.)*

(f) **Corollaries:**

  (*i*) *$\leq$ 'orders' the natural numbers. In this 'order':*

  (*ii*) *0 is the first element*

  (*iii*) *for each n, $n + 1$ is the immediate successor of n, and*

  (*iv*) *each $n \neq 0$ has an immediate predecessor.*

(g) *Put $W_n = \{m: m < n\}$. Then $\overline{\overline{W_n}} = n$. So every finite set A can be ordered (i.e., for some R, (A, R) is an order).*

Clearly we are reaching the point where many of the results given occur, with the *same* proofs, in books in algebra or analysis. So we give only a few more such propositions.

(h) *If (A, R) is any order, every finite subset of A has a first element (so also a last).*

Hence *every finite order is a well-order.*

(i) *Any two orders $(A, \leq)$, $(A', \leq')$ each on an n-element set (i.e., $\overline{\overline{A}} = n = \overline{\overline{A'}}$) are isomorphic.*

(j) *If $\kappa \leq^* n$ then $\kappa \leq n$.*

Finally, we add (to 2.3) the 'least element principle' and the dual, 'course-of-values induction'.

Proposition 2.4.

(a) *(The least element principle). If for some n, $\mathcal{P}\, n$, then there is a minimal (which is here the same as minimum) n such that $\mathcal{P}\, n$.*

(b) *(Course-of-values induction). If, for any n, if $\mathcal{Q}\, m$ holds for all $m < n$, then $\mathcal{Q}\, n$; then, for all n, $\mathcal{Q}\, n$.*

Proof of (a). Suppose $\mathcal{P}\, n$. If $\mathcal{P}\, m$ for no $m < n$, then n is minimal as desired. Otherwise $\{k \in W_n: \mathcal{P}\, k\}$ is non-empty, and so, being finite, has a least element m, by 2.3(h). It is easy to see that m is the least number with the property $\mathcal{P}$, as desired.

Proof of (b). Assume the hypothesis of (b) holds and that, for some n, $\mathcal{Q}n$ fails. By (a) let k be the least such n. Thus $\mathcal{Q}m$ holds for all $m < k$, so by our hypothesis, $\mathcal{Q}k$ holds, a contradiction.

Indeed, (b) not only follows from (a) but is almost identical with (a). The *contrapositive* of a statement having the form 'if $\mathcal{R}$ then $\mathcal{S}$' is the statement: "If not $\mathcal{S}$ then not $\mathcal{R}$'. Obviously such a statement and its contrapositive are always (logically) equivalent. Let us use also the obvious fact that '$\sim \forall x \Gamma x$' is logically the same as '$\exists x \sim \Gamma x$'. The reader (perhaps best using symbolic logic!) will now readily verify that (b) is in this way just exactly the contrapositive of (a), taking '$\mathcal{Q}$' to be '$\sim \mathcal{P}$'!

Notice that by (a) and 2.3(f), *if N exists, then $(N, \leq)$ is a well-order.*

Having now the notion of natural number, we can make up in set theory some more 'perfect copies' of old notions, as in Chapter 1. By an *n-termed sequence* or synonymously, *an ordered n-tuple of elements of A*, we mean a function s on W_n to A (or, in other words, a member of A^{W_n}). We may sometimes write s_i instead of $s(i)$. The *length* of s is n. We say t is a *finite sequence* if it is an n-termed sequence for some n. We say that R is an *n-ary relation* for A if $R \subseteq A^{W_n}$. o is called an n-ary *operation* for A if o: $A^{W_n} \to A$. (An ordered *2-tuple* is not exactly the same thing as an ordered couple, but many people just pretend that it is!)

Problems 14 and 15 are still some more well-known and useful facts about finite sets.

Problems

1. Prove 2.2(a) and (b).
2. Suppose we have a notion $\mathfrak{N}$ for which 2.2 can be proved, i.e., such that we can prove the following:
(a) $\mathfrak{N}0$. (b) If $\mathfrak{N}\kappa$ then $\mathfrak{N}(\kappa + 1)$. (c) (In general) If $\mathcal{P}0$ and for any λ such that $\mathfrak{N}\lambda$, if $\mathcal{P}\lambda$ then $\mathcal{P}(\lambda + 1)$, then for all λ such that $\mathfrak{N}\lambda$, $\mathcal{P}\lambda$. Prove that, for any λ, $\mathfrak{N}\lambda$ if and only if λ is a natural number.
3-12. Prove (respectively) 2.3(a), (b), (c), (d), (e), (f), (g), (h), (i), (j). Hint: Use induction except for 'Corollaries'. (These are very good problems for readers who have not already 'mastered' induction in other courses. Every reader should go through virtually the whole 3-12, at least in his head.)
13. Prove 2.4(b) (and hence its contrapositive (a)) in a different way using regular induction, i.e., 2.2(c) (and not 2.3(h)). Hint: take $\mathcal{P}n$ to be 'for all $m \leq n$, $\mathcal{Q}n$'.
14. Prove without AC the 'axiom of choice for finitely many sets': If I is finite, F is on I, and for each $i \in I$, $F(i) \neq \emptyset$ then there is a choice function for F.
15. Prove: If A is finite and f: $A \to A$ then f is one-to-one if and only if f is onto (critical in vector space theory).

§ 3. Multiplication and exponentiation

We return to arbitrary cardinals, for which we have so far defined + and $\leq$.

3.1 (***Definition of multiplication***).
Proposition (a). *If* $A \sim A'$ *and* $B \sim B'$ *then* $A \times B \sim A' \times B'$
Definition (b). $\kappa \cdot \lambda$ *(or just* $\kappa\lambda$*) = the unique* μ *such that, for some* $A, B, \overline{\overline{A}} = \kappa$, $\overline{\overline{B}} = \lambda$, *and* $\overline{\overline{A \times B}} = \mu$.

Proposition 3.2.
(a) $\kappa\lambda = \lambda\kappa$.
(b) $\kappa(\lambda\mu) = (\kappa\lambda)\mu$.
(c) $\kappa(\lambda + \mu) = \kappa\lambda + \kappa\mu$.
(d) $\kappa \cdot 0 = 0$; $\kappa \cdot 1 = \kappa$; $\kappa \cdot 2 = \kappa + \kappa$.
(e) *If* $\kappa\lambda = 0$ *then* $\kappa = 0$ *or* $\lambda = 0$.
(f) *If* $\kappa \leq \lambda$ *then* $\kappa\mu \leq \lambda\mu$.
(g) *If* $\kappa, \lambda \geq 2$ *then* $\kappa + \lambda \leq \kappa \cdot \lambda$.
(h) $m \cdot n$ *is a natural number.*
(i) *If* $m \neq 0$ *and* $n < p$ *then* $mn < mp$ *(so, if* $m \neq 0$ *and* $mn = mp$ *then* $n = p$*).*

The proofs of 3.1(a), and 3.2(a) - (e) are all direct (returning to sets from cardinals) and easy. (g) has a direct proof, and also one by algebra (see Problem 1). (f) and (i) are easy just by algebra. (h) is of course by an easy induction.

3.3 (***Definition of exponentiation***).

Proposition (a). *If* $A \sim A'$ *and* $B \sim B'$ *then* $A^B \sim A'^{B'}$.
Definition (b). κ^λ = *the unique* μ *such that for some* $A, B, \overline{\overline{A}} = \kappa$, $\overline{\overline{B}} = \lambda$, *and* $\mu = \overline{\overline{A^B}}$.

Proposition 3.4.
(a) $(\kappa^\lambda)^\mu = \kappa^{\lambda\mu}$.
(b) $\kappa^{\lambda + \mu} = \kappa^\lambda\kappa^\mu$.
(c) $(\kappa\lambda)^\mu = \kappa^\mu\lambda^\mu$.

Proof. For (a): We first easily see that (a) reduces to showing that for any A, B, C, $(A^B)^C \sim A^{B \times C}$. Thus we want to define F so that

$$(1) \qquad (A^B)^C \underset{F}{\sim} A^{B \times C}.$$

Without thought, we can say all of the following: Let F be the function on $(A^B)^C$ such that for any $f \in (A^B)^C$, $F(f)$ is the function on $B \times C$ such that for any $b \in B$ and $c \in C$, $F(f)(b, c)$ = ?! Now we have to think; but, in practice, only to the point of finding *some* combination of f, b, and c to put in the blank (as there will only be one reasonable possibility!) In our case we take:

$$(2) \qquad F(f)(b, c) = f(c)(b).$$

Now to show (1), we must go through a boring proof that F is into and onto $A^{B \times C}$ and one-to-one (see Problem 3). The proofs of (b) and (c) (Problems 4 and 5) are similar, but the 'boring parts' are omitted.

The kind of thing involved in proving 3.4(a), (b), and (c) (due to Cantor) plays a key role in various parts of modern mathematics, especially in functional analysis. When definitions like (2) above (of F) are made out of the blue they appear very mysterious and ingenious, unless (as Professor Dana Scott used to do) someone says "That's just $((\kappa^\lambda)^\mu = \kappa^{\lambda\mu})$"! In category theory there is even an attempt to single out maps like F above from just any old one-to-one onto map by a technically defined notion: 'F is a *natural* map'.

We add some very easy facts to 3.4.

Proposition 3.4 (continued).

(d) $\kappa^0 = 1$; $\kappa^1 = \kappa$; $\kappa^2 = \kappa\kappa$; *and if* $\kappa \neq 0$ *then* $0^\kappa = 0$.

(e) *If* $\kappa \leq \lambda$ *then* $\kappa^\mu \leq \kappa^\lambda$.

(f) *If* $\kappa \leq \lambda$ *then* $\mu^\kappa \leq \mu^\lambda$ *unless* $\kappa = \mu = 0 < \lambda$.

(g) *If* $\overline{\overline{A}} = \kappa$, *then* $\overline{\overline{P(A)}} = 2^\kappa$.

(h) $\kappa \leq 2^\kappa$.

(i) m^n *is a natural number.*

Proof. (d) and (e) are proved like (a)-(c) but very easily. (f) can be proved by algebra (Problem 6). (g) is the same as 1.6.1. (h) is trivial, mapping $a \in A$ into $\{a\} \in P(A)$. As usual, (i) is proved by an easy induction.

We do not give special laws of exponents for natural numbers, as they are proved in ordinary mathematics by the same methods we would use.

Soon (in § 3 of Chapter 4) we will prove Cantor's famous theorem that, for all κ, $\kappa < 2^\kappa$. In particular this result showed for the first time that there are two and indeed many different kinds of infinity (i.e., many different infinite cardinals). On the other hand, we will show (in § 4 of Chapter 4) that $\kappa + \kappa = \kappa \cdot \kappa = \kappa$ if κ is the cardinal number of N or of the set of reals; and (in Chapter 7) that the same holds for any infinite κ assuming AC. This may seem at first to make the arithmetic of infinite cardinals trivial, but it does so only partly. In fact the simple-minded laws like 3.4(a), etc., of this chapter must often be used together with the law $\kappa + \kappa = \kappa \cdot \kappa = \kappa$ in making calculations.

We now put off briefly the further study of arbitrary cardinals (until Chapter 4). In the rest of this chapter we study further the natural numbers, and in the very brief Chapter 3, we discuss how to introduce the integers, rationals, and reals.

Problems

1. Prove 3.2(g)
2. Prove 3.3(a).
3. Complete the proof of 3.4(a).

4. Prove 3.4(b), but only go as far as in the text for (a), i.e., define the required map like F in (a). (Do not show F is into, onto, or one-to-one.)
5. Prove 3.4(c) but only as far as in Problem 4.
6. Prove 3.4(f).
7. Improve (h) to: $2\kappa \leq 2^{\kappa}$.

§ 4. Definition by induction

In this section we study *recursion* (which means the same thing as *definition by induction*) over the natural numbers. In Chapter 7, more generally, recursion over an arbitrary well-ordering or over all ordinals is studied. The work in Chapter 7 is done without using the present §, which can therefore be omitted, *in theory*. But in practice it is easier for the reader to understand recursion first for the natural numbers. The general proof (in Chapter 7) is then seen as an easy modification of the one in this section. (This depends on the fact that we work in this section without the assumption that N exists.)

As a typical example of definition by induction, we consider the notion $\hat{n}$ (hyperexponentiation to the base 2 at n). We want to have $\hat{0} = 2$, $\hat{1} = 2^2$, $\hat{2} = 2^{2^2}$, $\hat{3} = 2^{2^{2^2}}$, and so on. We therefore say something like this: The notion $\hat{n}$ is defined inductively by requiring that

$$\begin{cases} \hat{0} = 2 & \text{and} \\ \widehat{(n+1)} = 2^{\hat{n}} & \text{for all } n. \end{cases}$$

Intuitively these conditions determine $\hat{n}$ for every n, because: $\hat{0} = 2$, so $\hat{1} = 2^{\hat{0}} = 2^2$; hence $\hat{2} = 2^{\hat{1}} = 2^{2^2}$; etc., etc.; and eventually one gets to any n. Thus the principle allowing us to make such definitions by induction (as in this example) appears to us intuitively to be valid. Indeed, mathematicians have made such definitions for hundreds of years.

Dedekind [F1888] made the key discovery that in fact this principle does not have to be taken as a new principle or assumption, but can be justified by other kinds of reasoning, which we are depending on anyway. This justification of recursive definitions will be the content of the following theorem 4.1, and its corollaries 4.2 and 4.3. Indeed, 4.1 (really 4.1(b)) gives a prescription by which an inductive definition like that of $\hat{n}$ above can be replaced by an ordinary explicit definition.

While in our example the values $\hat{n}$ are also natural numbers, this is often not the case.

In order to state and prove a result, 4.1(b), (c), about definition by induction over *all* natural numbers, we first prove a similar result, 4.1(a), about definition by induction over the numbers below a certain (arbitrary) number q.

Theorem 4.1. (Definition by induction).
(a) *There is exactly one f such that*

(1) f *is on* W_{q+1}, $f(0) = a$, *and* $f(n + 1) = \mathcal{A}_{f(n)}$ *for all* $n < q$.

Proof. Let us say that f works for q if (1) holds. First we show that:

(2) If f and g both work for q then $f = g$.

Since Dom f = Dom g, (2) follows from

(3) for any i, if $i \leq q$ then $f(i) = g(i)$.

(3) is easily proved by induction on 'i' (see Problem 1).

Now we show by induction on 'q' that (a) holds for all q. (Of course (a) now reduces to '*there exists an f which works for q*'.) Clearly $f = \{(0, a)\}$ works for 0. Assume g works for q. Then $f = g \cup \{(q + 1, \mathcal{A}_{g(q)})\}$ is seen almost at once to work for $q + 1$. Thus (a) is proved.

Since we are not assuming now the existence of the set of all natural numbers, the corresponding theorem 'over all natural numbers' must take a rather different *form* (though no new ideas are needed for the proof). The main question of formulation arises in saying the 'there exists' part, 4.1(c), which we therefore separate from the 'at most one' part, 4.1(b).

First we need a small lemma both parts of which the reader can verify in his head. (*No* induction is required.)

Lemma. (*i*) *If f works for n and $k < n$ then $f \restriction W_{k+1}$ works for k.*
(*ii*) *If*

(*) $\mathcal{B}_0 = a$ *and for all* n, $\mathcal{B}_{n+1} = \mathcal{A}_{\mathcal{B}_n}$,

then $(\mathcal{B}_i\colon i \leq k)$ *works for* k.

Theorem 4.1. (b) *If* (*) *holds for* $\mathcal{B}$ *and also for* $\mathcal{C}$ *then for all* n, $\mathcal{B}_n = \mathcal{C}_n$.

Proof. Again no induction is required. Since $(\mathcal{B}_i\colon i \leq n)$ and $(\mathcal{C}_i\colon i \leq n)$ both work for n (by the Lemma), they are equal by 4.1(a), so $\mathcal{B}_n = \mathcal{C}_n$.

It is easier to grasp the main result 4.1(c), when one considers a (typical) special case. If + is already known for the natural numbers, but $\cdot$ is not, then intuitively one can introduce $\cdot$ by recursion, saying:

(4) The notion $m \cdot n$ is defined by recursion on 'n' by requiring (for all m, n)

$$\text{(4)(a)} \begin{cases} m \cdot 0 = 0, \\ m \cdot (n + 1) = m \cdot n + m. \end{cases}$$

(This example differs from $\hat{n}$ above in having a parameter, '*m*'; but this causes no difficulty; just think of *m* as fixed). To justify saying (4) we need:

Theorem 4.1(c). (Special case). *Let us define $m \cdot n$ (in an ordinary, explicit way) by saying:*

$$(5) \quad \begin{cases} m \cdot n \text{ is the unique } z \text{ such that for some } f \text{ on } W_{n+1}, \\ f(0) = 0, f(i+1) = f(i) + m \text{ for all } i < n, \text{ and } f(n) = z. \end{cases}$$

Then we can prove that 4(a) holds.

Proof. (One can almost see this in his head.) In fact, using 4.1(a) and the Lemma above, a proof can be given making no use of induction. See Problem 2.

Once one can write down (5) himself, he has grasped this entire section!

Note that in practice (as in the statement (4)) we never give the explicit definition! This is because, if we have two notions $\cdot$ and $\cdot'$ introduced by two different explicit definitions, but for each we can prove the condition 4(a), then by 4.1(b) we can prove; for all *m* and *n*, $m \cdot n = m \cdot' n$.

Some readers may wish to skip the full statement 4.1(c) below. One thing there is new: It is seen that a certain variable-binding operator '$D_x(\mathfrak{A}_x, a, n)$' – like '$\bigcup_{i \in I} \mathfrak{A}_i$' – naturally occurs here. (The author first heard of this from work of A. P. Morse.)

4.1 (c). ***Definition.*** $D_x(\mathfrak{A}_x, a, n) =$ *the unique z such that for some f on* W_{n+1}, $f(0) = a$, $f(i+1) = \mathfrak{A}_{f(i)}$ *for all* $i < n$, *and* $f(n) = z$.

Theorem. $D_x(\mathfrak{A}_x, a, 0) = a.$
$D_x(\mathfrak{A}_x, a, n+1) = \mathfrak{A}_{D_x(\mathfrak{A}_x,\, a,\, n)}.$

The proof of 4.1(c) is the same as that for the special case (cf. Problem 2).

Note that a result, like the Special Case, which has a parameter follows from 4.1(c) for this reason: In the usual, informal logic, any theorem involving '$\mathfrak{A}$' is considered to remain valid when one replaces each occurrence of $\mathfrak{A}_x$($\mathfrak{A}_y$, etc.) by a certain naming expression *with parameter* – like $x + m$. (Of course one replaces $\mathfrak{A}_y$ by $y + m$, etc.)

Naturally there is also a course-of-values form of recursion, which is in 4.2 below. Notice that in defining $\mathfrak{B}_n$ for *n*, we now are to know what $\mathfrak{B}_i$ is for each $i < n$. The mathematical entity which has exactly that information content is $(\mathfrak{B}_i : i < n)$. We state 4.2 informally.

Corollary 4.2. *We can justifiably introduce by recursion a notion $\mathfrak{B}$ by requiring that*

$$(6) \quad \mathfrak{B}_n = \mathfrak{C}((\mathfrak{B}_j : j < n)) \text{ for all } n.$$

Proof. There are two proofs. In one (which we ignore), one repeats with suitable adjustments the whole 4.1(a), (b), (c) and its proof. In the other, we can obtain 4.2 as a direct corollary of 4.1. We give some hints for this proof and let the reader complete it in Problem 3. Hints (*rough*): Define, by 4.1, a notion $\mathcal{O}_n$ which is intended to satisfy the equation $\mathcal{O}_n = (\mathcal{B}_i : i < n)$ – when we finally get $\mathcal{B}$.

We have at once from 4.1 (or 4.2):

Corollary 4.3. *If N exists, then in fact there is exactly one* function *F on N such that* $F(0) = c$ *and, for all n,* $F(n + 1) = \mathcal{A}_{f(n)}$. *(Similarly for course-of-values.)*

Problems

1. Prove (3).
2. Prove 4.1(c) (Special Case).
3. Obtain 4.2 as a corollary of 4.1. (See hints there.)

§ 5. Axiom of infinity, Peano axioms, Dedekind infinite sets

An algebraic structure (A, S, z) – where $z \in A$ and $S: A \to A$ – is called a *Peano structure* if (P1) for any $a \in A$, $Sa \neq z$; (P2) S is one-to-one; and (P3) for any subset B of A, if $z \in B$ and $Sa \in B$ whenever $a \in B$, then $B = A$. (These are the famous Peano axioms.) We take for the Axiom of Infinity the statement:

Axiom of Infinity. *There exists a Peano structure.*

Theorem 5.1. *The following are equivalent (forms of the Axiom of Infinity):*
- (a) *There exists a Peano structure;*
- (b) *There exists an infinite set;*
- (c) *N exists.*

Proof. If N exists and Sc is the successor function on N ($Sc(n) = n + 1$) then from § 2 we see at once that (N, Sc, z) is a Peano structure. Thus (c) implies (a). In any Peano structure (A, S, z), A is (by means of S) equivalent to a proper subset of itself; so by § 2, A is infinite. Thus (a) implies (b). If we assume (b) then some infinite κ exists. By 1.7, $\{\lambda : \lambda \leq \kappa\}$ exists, so clearly the set $W = \{n : n \leq \kappa\}$ exists. But by 2.3(e), $n \leq \kappa$ for every n. So $N = W$ and N exists. Thus (b) implies (c).

Notice that *form* (b) *justifies the name 'axiom of infinity'*. (We take the axiom to be (a) officially, as (a) only involves $\in$, while at this point, 'infinite' involves $=$ as well.) In this connection, notice that the Axiom of Infinity does not say that there are infinitely many things (sets). Indeed, without it, we have the infinitely many sets $\emptyset$, $\{\emptyset\}$, $\{\{\emptyset\}\}$, $\{\{\{\emptyset\}\}\}$, $\cdots$. It is just with the existence of a single set which is infinite that the typical and sometimes bizarre character of set theory begins.

We now adopt the Axiom of Infinity in the rest of Chapter 2 and Chapters 3-5. We put $\aleph_0 = \overline{\overline{N}}$. ('$\aleph$' is the first letter, aleph, of the Hebrew alphabet.)

If we use the Axiom of Choice, we obtain (in 5.2) a result which goes beyond (b) $\rightarrow$ (c). Incidentally, the proofs of 5.2 and 5.3 both offer good examples of the use of recursive definitions.

Theorem 5.2^{AC}. *If κ is infinite, then $\aleph_0 \leq \kappa$.*

Proof. Let $\overline{\overline{A}} = \kappa$. Since A is not empty, we can choose $a_0 \in A$. If we have chosen distinct $a_0, \ldots, a_n \in A$, then, since A is infinite, $A - \{a_0, \ldots, a_n\}$ is not empty so we can pick $a_{n+1} \in A$ different from each of $a_0, \ldots, a_n$. Then clearly $N \sim \{a_n : n \in N\} \subseteq A$ as desired.

The kind of argument just given is used by most mathematicians, since it seems to be easier to find than a more 'proper' argument. The reader should probably employ such arguments himself and at the same time know how to replace them by arguments using only the Axiom of Choice as stated. In this case, let F be a choice function for $P(A) - \{\emptyset\}$. Define a_n by course-of-values recursion (4.2) by requiring that, for each n, $a_n = F(A - \{a_i : i < n\})$. The rest is as before.

We now show that the Peano axioms *characterize* $(N, Sc, 0)$ *up to isomorphism*:

Theorem 5.3. *(A, S, z) is a Peano structure if and only if $(A, S, z) \cong (N, Sc, 0)$.*

Proof. $(\leftarrow)$ follows at once from results in § 2. Assume (A, S, z) is a Peano structure. By 4.3 we can take f to be the function on N such that

$$\begin{cases} f(0) = z, & \text{and} \\ f(n+1) = S(f(n)) & \text{for all } n. \end{cases}$$

We see at once by induction on 'n' that each $f(n) \in A$. By applying (P3) to the right B one easily shows (a) f is onto A (Problem 1). Once can also show (b) $f(m) \neq f(n)$ whenever $m \neq n$ (Problem 2). It follows that $(N, Sc, 0) \underset{f}{\cong} (A, S, z)$, since f 'preserves' S and 0 by its defining conditions. Thus 5.3 is proved.

A cardinal κ is called *Dedekind infinite* if $\kappa \geq \aleph_0$. Thus Dedekind infinite implies infinite, and given *AC* the converse holds, by 5.2. For some more study (in the absence of *AC*) of 'Dedekind infinite', see Problem 3.

Problems

1. Prove (a) in the proof of 5.3.
2. Prove (b) in the proof of 5.3.

3. (*AC* is *not* assumed.) Prove that the following are equivalent: (a) κ is Dedekind infinite, i.e., $\aleph_0 \leq \kappa$; (b) $\kappa = \kappa + 1$; (c) for some A, $\kappa = \overline{\overline{A}}$ and A is equivalent to a proper subset of itself. Hints: For (a) $\rightarrow$ (b) write $\kappa = \mu + \aleph_0 = \mu + \aleph_0 + 1$. (b) $\rightarrow$ (c) is trivial. The main thing is (c) $\rightarrow$ (a). Here one must make up a specific set of power $\aleph_0$, given that $A \underset{f}{\sim} B \subseteq A - \{a\}$.

3. THE NUMBER SYSTEMS

§ 1. Introductory remarks

$(N, Sc, 0)$ was characterized up to isomorphism as a Peano system in 2.5.3. This characterization can be very easily modified to apply to the structure $(N, +, \cdot, 0, 1, <)$ which we call $\underline{N}$. Indeed, let us say that $\underline{A} = (A, +, \cdot, 0, 1, <)$ is a *Peano semiring* if: $(A, S, 0)$ is a Peano system, where $Sa = a + 1$ for any $a \in A$; for any $a, b \in A$, $a < b$ if and only if for some $c \neq 0$, $b = a + c$, and also

$$\begin{cases} a + 0 = a \\ a + (b + 1) = (a + b) + 1 \end{cases} \text{and} \begin{cases} a \cdot 0 = 0 \\ a \cdot (b + 1) = a \cdot b + a. \end{cases}$$

Using various facts from Chapter 2, the reader can easily check that

1.1 *$\underline{A}$ is a Peano semiring if and only if $\underline{A}$ is isomorphic to* $\underline{N} = (N, +, \cdot, 0, 1, <)$.

It follows that *any* Peano semiring obeys the following rich list of laws (which should be taken as restricted to the natural numbers): 1.3, 1.4(a), (c), 2.3(c), (d), (f), (g), (h), (i), and 3.2 (all from Chapter 2). (These include, for example, $(m \cdot n) \cdot p = m \cdot (n \cdot p)$, etc., etc.)

In many books on either modern algebra or modern analysis, the book begins by taking as given a structure $(N, +, \cdot, 0, 1, <)$, which is assumed to satisfy certain postulates. These postulates always include being a Peano semiring. They usually include also many other laws, *all of which will be found in our list above.* (We have deliberately arranged this in Chapter 2). Such books then embark on a discussion of the construction and characterization up to isomorphism of the integers, rationals, and reals. By what was just said above, these discussions can be taken *verbatim* to lie within our development (right at this point).

Because of this state of affairs it is optional whether or not to include in a book like this one a full discussion of the construction of the various number systems. We have chosen not to do so, because such a discussion interrupts too much the natural flow or development of Cantor's set theory. However, we shall now in § 2 state precisely the three main theorems of the subject. One purpose of this is to ensure that the reader sees that the whole subject, including treatments in books on algebra or analysis, can be considered to take place within our development, and, indeed, just after Chapter 2.

§ 2. Construction and characterization up to isomorphism of the integers, rationals, and reals

We shall use a number of familiar notions like 'group' 'ordered field', etc., etc., whose definitions are given early in any book on algebra. Also we assume very simple algebraic facts, such as that in a group $(A, +)$ we can introduce the notion $-a$.

If $\underline{A} = (A, +, \cdot, 0, 1, <)$ is an ordered integral domain and $a \in A$, we define (as usual) the notion $n \circ a$ by recursion, specifying that $0 \circ a = 0$ and $(n + 1) \circ a = n \circ a + a$, for any $a \in A$.

Theorem 2.1. *There exists one and up to isomorphism only one ordered integral domain* $\underline{Z} = (Z, +, \cdot, 0, 1, <)$ *such that every element* x *of* Z *is of the form* $n \circ 1$ *or* $-(n \circ 1)$ *(where* $n \in N$*).*

Remarks on the proof. Showing that any two such $\underline{Z}$'s are isomorphic is a routine problem (Problem 1) using simple algebra and 1.1. Now we consider how to construct such a $\underline{Z}$.

The most elementary way (indeed the way used in elementary school!) is to add to N *new* numbers "$-n$" (say, take $-n = (1, n)$) for $n \neq 0$. However, the details are a little shorter if instead we consider all ordered couples (m, n) (intuitively thought of as $m - n$), introduce an equivalence relation ($(m, n) \approx (m', n')$ if and only if $m + n' = m' + n$) and take Z to be the set of all cosets $(m, n) / \approx$ – much as in the better known proof of 2.2 below. (Of course we must still define the $+$, $\cdot$, and $<$ of $\underline{Z}$. But the method we should try to use is clear.)

Sometimes it is convenient to ask a little more than is stated in 2.1. In fact, in either construction above *it is possible* to *define* a definite Z, (henceforth the only thing called $\underline{Z}$) *and secondly to arrange that our original* $\underline{N}$ *is directly embedded in* $\underline{Z}$.

If $(A, +, \cdot, 0, 1, <)$ is an ordered integral domain, $a \in A$, and $i = n$ or $i = -n$ we put (as usual) $i \circ a = n \circ a$ or $-(n \circ a)$, respectively.

Theorem 2.2. *There is one and up to isomorphism only one ordered field $\underline{Q} = (Q, +, \cdot, 0, 1, <)$ such that every element of Q is of the form $(i \circ 1)/(j \circ 1)$ (where $i, j \in Z$ and $j \neq 0$).*

This is the well-known 'field of quotients' which can be formed over any integral domain (in the role of $\underline{Z}$ in 2.2.). These details are in every modern algebra book. The uniqueness up to isomorphism part is Problem 2.

Again, *one can define a certain definite $\underline{Q}$ which also embeds $\underline{Z}$* (i.e., for any $i \in Z$, $i \circ 1 = i$).

By a *complete* ordered field is meant an ordered field $\underline{A} = (A, +, \cdot, 0, 1, <)$ in which every non-empty subset bounded above has a least upper bound.

Theorem 2.3. *There is one and up to isomorphism only one complete ordered field $\underline{R} = (R, +, \cdot, 0, 1, <)$.*

Remarks about the proof. For uniqueness up to isomorphism, see Problem 3. The two best known methods of construction are those of Dedekind and Cantor. In Dedekind's, we take R to be the family of all non-empty proper initial segments of $(Q, <)$ having no largest element. In Cantor's we consider infinite sequences x of rationals (i.e., $x : N \to Q$) which are 'Cauchy' sequences, i.e., for any k there exists m such that for all $n \geq m$, $|x_n - x_m| < \frac{1}{k}$. A certain equivalence relation E ('converging together') is introduced among such x's, and R is taken to be the set of all cosets x/E. With these hints the proofs in both moethods are fairly straightforward, though laborious. The details can be found in many books on algebra, analysis, the number systems, or set theory – or, least boringly, worked out by oneself.

Again *we can in addition define a definite R and one which directly embeds $\underline{Q}$*.

Notice that the proofs for 2.1 and 2.2 are close to things one learns in grammar school, but neither of the two proofs discussed for Theorem 2.3 is. Actually there is a third proof constructing say the positive reals to be like decimals $d_1 d_2 \cdots d_k . d'_1 d'_2 \cdots d'_n \cdots$. This is, of course, close to school mathematics. However, the details for the third proof are quite tedious, so this 'least sophisticated' proof is rarely given in full.

Remarks. Of course, 2.1 – 2.3 are additional examples (the most famous) of the process of absorbing old mathematical notions within set theory. The main steps of this kind are now almost at an end (except for the *definition* of cardinals, etc., in Chapter 7). Indeed, nearly all of mathematics (for example, courses or books on algebraic geometry, differential topology, etc., etc.) can be considered, *verbatim*, to be taking place in our set theory to date (assuming 2.1 – 2.3 have been proved.) Hence, we are now entitled to assume (and indeed a few times in Chapter 4 we *shall* assume) knowledge of various elementary facts from number theory, algebra, or analysis.

We have assumed the Axiom of Infinity throughout Chapter 3 but it is essential only in constructing the reals. In 2.2, for example, in order to avoid the Axiom of Infinity, we would take m/n to be (k, ℓ) where $\ell > 0$ and 'k/ℓ is m/n put in lowest terms'. (The equivalence class ordinarily used is infinite).

We already mentioned that the construction of the rationals holds in a broader context. Each of the two main constructions of the reals can also be carried out almost unchanged in a certain important, broader context or contexts. Dedekind's method produces a result about 'completing' any partial ordering. Cantor's construction has two important generalizations. One is in analysis and is about the 'completion' of an arbitrary metric space. The other is in algebra or number theory, where the method is used to construct the important *field of p-adic numbers* (p any fixed prime). One may as well verify one of these generalizations when he verifies the special case in 2.3, as the proofs are essentially the same. See for example van der Waerden [F1949].

Problems

1. Prove 'uniqueness up to isomorphism' in 2.1.
2. The same for 2.2.
3. The same for 2.3. Hint: To apply the result of Problem 2, construct the set W of elements $\frac{i \cdot 1}{j \cdot 1}$ $(i, j \in Z, j \neq 0)$, show W is isomorphic to $\underline{Q}$ and W is dense in R. For the latter, first show $\{n \circ 1 : n \in N\}$ has no upper bound (i.e., $\underline{R}$ is 'Archimedian').

4. MORE ON CARDINAL NUMBERS

Having been sidetracked by discussions of the finite cardinals and the real numbers, we now return to Cantor's theory of arbitrary cardinal numbers.

§ 1. The Cantor-Bernstein Theorem

Theorem 1.1. (Cantor-Bernstein). *If* $\kappa \le \lambda$ *and* $\lambda \le \kappa$ *then* $\kappa = \lambda$.

This result was established by Cantor using the Axiom of Choice, and then in 1897 by F. Bernstein without the aid of the Axiom of Choice. In Chapter 7, Theorem 5.3., using *AC*, we will infer 1.1 in a few lines from other results needed anyway – thus giving Cantor's proof of 1.1.

In this section we will give the rather beautiful proof of 1.1 not depending on *AC*. But the reader may prefer just to assume 1.1 now, and wait for the easy proof in Chapter 7. There is no absolutely telling argument against this, since we do eventually adopt *AC*! Of course these remarks do not apply to a reader who strongly rejects the Axiom of Choice (and thus the key parts of this book!)

Perhaps any reader can at least look at the very nice pictures in the proof below:

Proof **(*of 1.1 without AC*).** Let $f: A \to B$ and $g: B \to A$ be one-to-one.

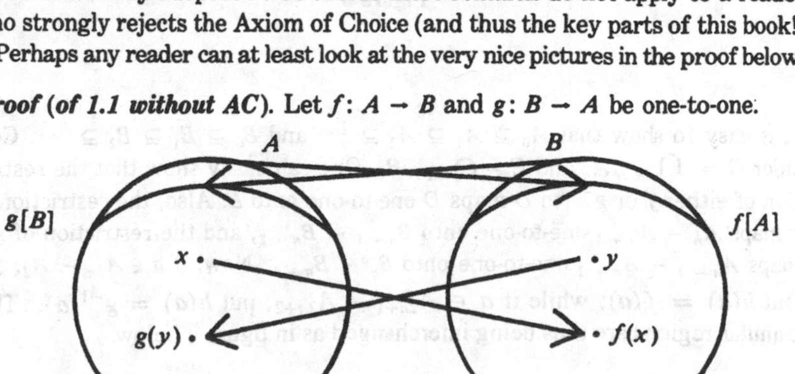

Figure 5

We shall *define* h such that $A \underset{h}{\sim} B$. (The AC proof gives only the pure existence of such an h.)

Define (A_n, B_n), $n = 0, 1, \cdots$ by recursion, specifying

$$\begin{cases} A_0 = A, & B_0 = B, \\ A_{n+1} = g[B_n], & B_{n+1} = f[A_n]. \end{cases}$$

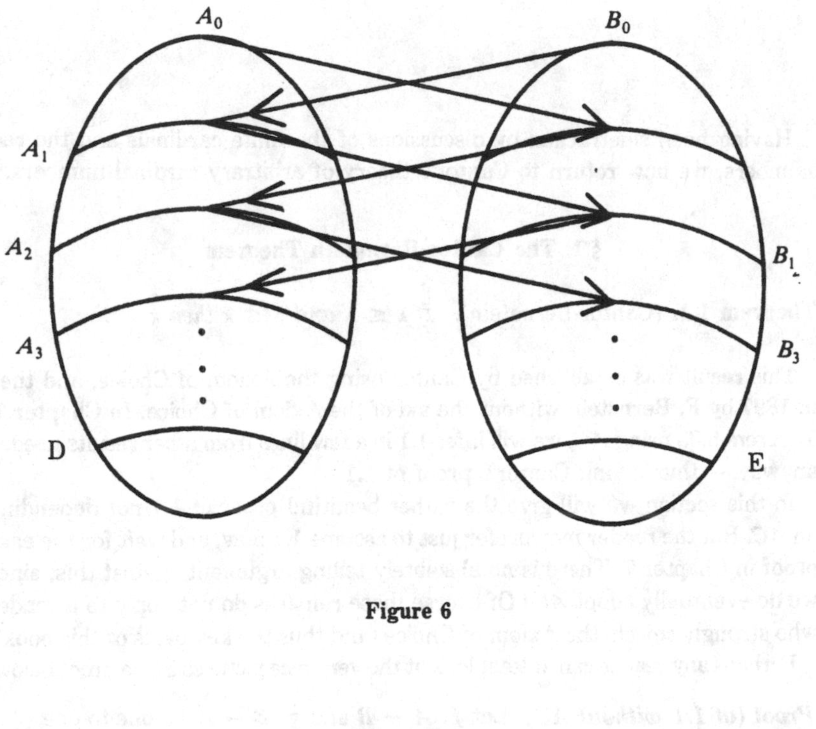

Figure 6

It is easy to show that $A_0 \supseteq A_1 \supseteq A_2 \supseteq \cdots$ and $B_0 \supseteq B_1 \supseteq B_2 \supseteq \cdots$. Consider $D = \bigcap_{n \in N} A_n$ and $E = \bigcap_{n \in N} B_n$. One can easily show that the restriction of either f or g^{-1} to D maps D one-to-one onto E. Also, the restriction of f maps $A_n - A_{n+1}$ one-to-one onto $B_{n+1} - B_{n+2}$, and the restriction of g^{-1} maps $A_{n+1} - A_{n+2}$ one-to-one onto $B_n - B_{n+1}$. Now, if $a \in A_{2n} - A_{2n+1}$, put $h(a) = f(a)$; while if $a \in A_{2n+1} - A_{2n+2}$, put $h(a) = g^{-1}(a)$. The annular regions are thus being interchanged as in figure 7 below.

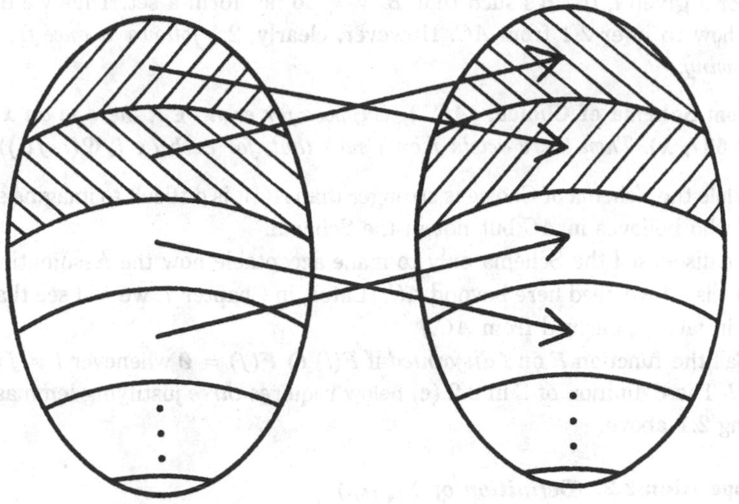

Figure 7

It is now easy to see that $A \underset{h}{\sim} B$.

Problem

1. Rewrite the proof, filling in all missing details.

§ 2. Infinite sums and products of cardinals

Cantor's addition and multiplication of cardinals (from Chapter 2) extend at once to notions of arbitrary infinite sums and products of cardinals. Cantor's infinite sums and products always exist. Moreover they are seen at once to obey all imagineable generalizations of the associative (distributive, etc.) laws! (In these respects, they differ greatly from the infinite sums (series) and products of analysis.)

All these things (proved below) depend essentially on the *Axiom of Choice, which is assumed throughout* § 2. Indeed, we make in § 2 also the following assumption:

Assumption 2.1. *Given* $(\kappa_i\colon i \in I)$, *there exists* F *on* I *such that* $\overline{\overline{F(i)}} = \kappa_i$ *for each* $i \in I$.

For a given i, the B's such that $\overline{\overline{B}} = \kappa_i$ do not form a set. Thus we do not see how to infer 2.1 from AC. However, clearly, 2.1 *follows at once from the following:*

Axiom Schema of Choice. (AC^+) *Suppose for each $i \in I$, there is an x such that $\mathcal{P}(i, x)$. Then there exists f on I such that, for each $i \in I$, $\mathcal{P}(i, f(i))$.*

While the Schema of Choice is stronger than AC it is difficult to imagine someone who believes in AC but not in the Schema.

We discussed the Schema only to make acceptable now the Assumption 2.1 which is all we need here beyond AC. (Later, in Chapter 7, we will see that 2.1 can in fact be derived from AC.)

Call the function F on I *disjointed* if $F(i) \cap F(j) = \emptyset$ whenever $i \neq j$ and $i, j \in I$. The definition of Σ in 2.2.(c) below requires *three* justifying lemmas, one being 2.1 above.

Proposition 2.2. ***(Definition of*** $\Sigma_{i \in I}\kappa_i$.)

(a) *If F and G are both disjointed and on I and for each $i \in I$, $F(i) \sim G(i)$, then $\bigcup_{i \in I}F(i) \sim \bigcup_{i \in I}G(i)$. (See Problem 1.)*

(b) *For any F* on *I there exists a disjointed G on I such that for each $i \in I$, $F(i) \sim G(i)$. (Put $G(i) = \{i\} \times F(i)$.*

Definition (c). *$\Sigma_{i \in I}\kappa_i$ is the unique μ such that there exists a disjointed F on I such that, for each $i \in I$, $\overline{\overline{F(i)}} = \kappa_i$, and $\mu = \overline{\overline{\bigcup_{i \in I} F(i)}}$. (See Problem 2).*

Similarly we have

2.3 (Definition of $\Pi_{i \in I}\kappa_i$).

(a) *If F and G are on I and for each $i \in I$, $F(i) \sim G(i)$, then $\Pi_{i \in I}F(i) \sim \Pi_{i \in I}G(i)$. (See Problem 3.)*

(b) ***Definition*** *(justified by (a) and 2.1). $\Pi_{i \in I}\kappa_i$ is the unique μ such that there exists F on I such that for each $i \in I$, $\overline{\overline{F(i)}} = \kappa_i$, and $\overline{\overline{\Pi_{i \in I}F(i)}} = \mu$.*

(Some claim that the Π in $\Pi_{i \in I}\kappa_i$ is smaller than the one in $\Pi_{i \in I}A_i$.)

The following laws look formidable at first glance, but are really easy to comprehend and even to prove. The proofs for 2.4 are left to the reader.

Theorem 2.4.

(a) *(General commutative law). Let $I \underset{f}{\sim} J$ (perhaps $I = J$). Then $\Sigma_{j \in J}\kappa_j = \Sigma_{i \in I}\kappa_{f(i)}$. (Likewise for Π.)*

(b) *(General associative law).*

$$\sum_{i \in I}\sum_{j \in J_i}\kappa_{ij} = \sum_{(i,j) \in W}\kappa_{ij}, \text{ where}$$

$W = \{(i, j)$: $i \in I$ and $j \in J_i\}$. (Likewise for Π.)

(c) *(General distributive law). $\kappa \cdot \Sigma_{i \in I}\lambda_i = \Sigma_{i \in I}(\kappa\lambda_i)$; indeed:*

$$\prod_{i \in I}\sum_{j \in J_i}\kappa_{ij} = \sum_{h \in \prod_{i \in I} J_i}\prod_{i \in I}\kappa_{ih(i)} \qquad \textit{(by 1.6.3).}$$

(d) $\Sigma_{i\in\{0,1\}}\kappa_i = \kappa_0 + \kappa_1$ *and* $\Pi_{i\in\{0,1\}}\kappa_i = \kappa_0 \cdot \kappa_1$. *Also,* $\Sigma_{i\in I}\kappa = \kappa\cdot\overline{\overline{I}}$ *and* $\Pi_{i\in I}\kappa = \kappa^{\overline{\overline{I}}}$.

Thus $1 + 1 + 1 + \cdots =$ (by definition) $\Sigma_{n\in N} 1 = \aleph_0$ and similarly $2\cdot 2\cdot 2\cdot\cdots = 2^{\aleph_0}$.

(e) *If* $J \subseteq I$ *and* $\kappa_i = 0$ *for all* $i \in I - J$, *then* $\Sigma_{i\in I}\kappa_i = \Sigma_{i\in J}\kappa_i$; *and likewise for* 1 *and* Π.

The results below in 2.5 can be obtained as corollaries of laws in 2.4 (or directly). They are left to the reader.

Theorem 2.5.
(a) *If, for all* $i \in I$, $\kappa_i \le \lambda_i$, *then* $\Sigma_{i\in I}\kappa_i \le \Sigma_{i\in I}\lambda_i$ *and* $\Pi_{i\in I}\kappa_i \le \Pi_{i\in I}\lambda_i$.
(b) *If* $\lambda \in Q$ *then* $\lambda \le \Sigma_{\kappa\in Q}\kappa$. *(Hence every set of cardinals is bounded above.)*
(c) $\kappa^{\Sigma_{i\in I}\lambda_i} = \Pi_{i\in I}\kappa^{\lambda_i}$.
(d) $(\Pi_{i\in I}\kappa_i)^\lambda = \Pi_{i\in I}\kappa_i^\lambda$. *(Note (d) includes both old laws* $(\kappa^\lambda)^\mu = \kappa^{\lambda\mu}$ *and* $(\kappa\lambda)^\mu = \kappa^\mu\lambda^\mu$.)

Proposition 2.6. *If* $\kappa_i \ge 2$ *for all* $i \in I$, *then* $\Sigma_{i\in I}\kappa_i \le \Pi_{i\in I}\kappa_i$ *(See Problem 4.)*

Problems
1. Prove 2.2(a) (*AC* is needed, but 2.1 is not.)
2. Check that Definition 2.1(c) is justified, i.e., there is always exactly one such μ.
3. Prove 2.3(a).
4. Prove 2.6. (Compare Problem 1 of § 3 of Chapter 2.)

§ 3. Different kinds of infinity

3.1 Cantor's Theorem. $\kappa < 2^\kappa$.

Proof. We saw easily in 2.3.4(h) that $\kappa \le 2^\kappa$; so we only need to show $\kappa \ne 2^\kappa$. Let $\overline{\overline{A}} = \kappa$ (so that $\overline{\overline{P(A)}} = 2^\kappa$). Suppose $B\colon A \to P(A)$, so that $(B_a\colon a \in A)$ is a 'list' of subsets of A. It will be enough to find some subset of A which is not any of the sets B_a (for $a \in A$). In fact, we can form a subset C of A which, for each $a \in A$, differs from the 'a-th set' B_a regarding the membership of a. To do this just put $C = \{a \in A\colon a \notin B_a\}$. Then at once $a \in C$ if and only if $a \notin B_a$. Thus C differs from each B_a, as desired. This shows that $A \not\sim P(A)$ so $\kappa \ne 2^\kappa$.

Aside. In fact, our argument showed that B cannot be onto even when it is not one-to-one. Thus we showed that $2^\kappa \not\le^* \kappa$, which is stronger than 3.1 in the absence of *AC*.

This argument is often called Cantor's *diagonal argument.* To see why, consider the case $A = N$, and consider $\{0, 1\}^N$ rather than the equivalent $P(N)$. If $f \in \{0, 1\}^N$ we can picture it as an infinite sequence or list of 0's and 1's:

$$f(0), f(1), f(2), \cdots, f(k), \cdots.$$

Now suppose $(f_n : n \in N)$ is a list of such f's. Can they be all? We have the picture

$$\begin{array}{ccccc} f_0(0) & f_0(1) & f_0(2) & f_0(3) & \cdots \\ f_1(0) & f_1(1) & f_1(2) & f_1(3) & \cdots \\ f_2(0) & f_2(1) & f_2(2) & f_2(3) & \cdots \\ f_3(0) & f_3(1) & f_3(2) & f_3(3) & \cdots \\ \vdots & \vdots & \vdots & \vdots & \end{array}$$

The new g we form has $g(n) = 1 - f_n(n)$ and so differs from f_n at the diagonal position pictured.

By the Cantor-Bernstein theorem, $<$ is transitive (check!) So from 3.1 we see at once that all of the cardinals $\aleph_0, 2^{\aleph_0}, 2^{2^{\aleph_0}}, \cdots$ are distinct. More precisely, and switching to sets rather than cardinals, we can define the sets $(B_n : n \in N)$ by recursion taking $B_0 = N$, and for each n, $B_{n+1} = P(B_n)$. Then $\overline{\overline{B}}_0 < \overline{\overline{B}}_1 < \overline{\overline{B}}_2 < \cdots$. But we are not through as the set $\bigcup_{n \in N} B_n$ clearly has power greater than each $\overline{\overline{B}}_k$. To analyze (even assuming AC) just how these cardinals go on and on we need the theory of well-orderings (see Chapter 7). But it is clear now that there are a great many infinite cardinals.

The next theorem is an important generalization of Cantor's Theorem 3.1, and is one of the few basic theorems of cardinal arithmetic not discovered by Cantor himself. It was found by J. König [F1905] for the case $\overline{\overline{I}} = \aleph_0$, and extended to the general case by Zermelo and Jourdain. It is traditionally called

3.2. The König-Zermelo Theorem. *(Assume AC and Assumption 2.1.) If, for each $i \in I$, $\kappa_i < \lambda_i$, then $\Sigma_{i \in I}\kappa_i < \Pi_{i \in I}\lambda_i$.*

First note that $\leq$ holds in 3.2. Indeed, by 2.4(e), leaving out 0's and 1's we can assume all $\lambda_i \geq 2$; and now $\leq$ follows from 2.6. The proof of $\neq$ in 3.2 (Problem 1) is like that of 3.1, but different enough so the reader can be quite pleased when he finds it!

Incidentally, to see that 3.2 implies 3.1 (assuming AC and 2.1), let $\kappa = \overline{\overline{A}}$. Then

$$\kappa = \sum_{a \in A} 1 < \prod_{a \in A} 2 = 2^{\kappa}.$$

An example of an important application of the König-Zermelo theorem is in § 1 of Chapter 11 (which can be read after Chapter 7).

Problems

1. Prove $\neq$ in 3.2.

§ 4. $\aleph_0$, $2^{\aleph_0}$, and $2^{2^{\aleph_0}}$ – the simplest infinite cardinals

A set A is called *denumerable* if $\overline{\overline{A}} = \aleph_0$, and *countable* if $\overline{\overline{A}} \leq \aleph_0$. (Thus, by definition, *every subset of a countable set is countable.)*

Theorem 4.1. *A is countable if and only if A is denumerable or finite.*

Proof. ($\leftarrow$) is obvious. *First proof* of ($\rightarrow$), *using AC*: Let $\kappa = \overline{\overline{A}} \leq \aleph_0$. If κ is not finite then by 2.5.2$^{(AC)}$, $\aleph_0 \leq \kappa$; so by the Cantor-Bernstein Theorem, $\kappa = \aleph_0$, as desired. *Second proof* avoiding *AC*: We can assume A is an infinite subset of N. By course-of-values recursion, we obtain F on N so that for each n, $F(n)$ is the $\leq$-first member of $A - \{F(k): k < n\}$. (This set is not empty as A is infinite.) Clearly F is one-to-one on N into A. So $\aleph_0 \leq \overline{\overline{A}}$ and $\aleph_0 = \overline{\overline{A}}$, as desired, by the Cantor-Bernstein Theorem. (The latter is easily avoided by showing F is onto.)

Proposition 4.2. *If $\kappa \leq^* \aleph_0$, then κ is countable. (Any image of a countable set is countable.)*

Proof. This is a special case of 2.1.8(c), since $\leq$ well-orders N.

Let us write $Sq = \bigcup_{n \in N} N^{W_n}$ = the set of all finite sequences of natural numbers.

In the next theorem we learn (with Cantor) that many sets of numbers such as Q, which appear to be more numerous than N, are not!

Theorem 4.3.

(a) *The sets Z (of integers) and Q (of rationals) are denumerable. Also:*

(1) $$\aleph_0 = \aleph_0 + \aleph_0 = \aleph_0 \cdot n = \aleph_0 \cdot \aleph_0 = \aleph_0^n = \overline{\overline{Sq}}.$$

Proof. There are many proofs. To save time we use facts about prime numbers, but they could be avoided.

In (1), each $\leq$ is obvious. By the Cantor-Bernstein Theorem (in a typical use) the whole (1) will follow from: $\overline{\overline{Sq}} \leq \overline{\overline{N}}$. To see this, if we are given $z = (k_0, \ldots, k_{n-1}) \in Sq$, put $f(z) = 2^{k_0 + 1} 3^{k_1 + 1} \cdots p_{n-1}^{k_{n-1} + 1}$, where $p_0, p_1, \ldots$ are the primes in order. Clearly f maps Sq one-to-one into N, which proves $\overline{\overline{Sq}} \leq \overline{\overline{N}}$, and hence (1). The sets Z and Q are easily seen to be equivalent to subsets or images of sets denumerable by (1). So (by 4.2, etc.) $\overline{\overline{Z}} = \overline{\overline{Q}} = \aleph_0$ (as clearly $\aleph_0 \leq \overline{\overline{Z}}, \overline{\overline{Q}}$).

A real number is called *algebraic* if it is a root of some polynomial with rational coefficients not all zero. Continuing with 4.3., we have

(b) *The set Alg of all real algebraic numbers is countable.*

Proof. Let K be the set of all finite sequences $a = (a_0, \ldots, a_n)$ of rationals where $a_0 \neq 0$. Thus a real number x is algebraic if and only if for some $a = (a_0, \ldots, a_n) \in K$

$$(2) \qquad a_0 x^n + a_1 x^{n-1} + \cdots + a_{n-1} x + a_n = 0.$$

By elementary algebra, such an equation has at most n roots. Say (2) has the distinct real roots $u_0, \ldots, u_{k-1}$ listed in their *real* order. For any a, k, as above, and any i, put $f(a, i) = u_i$ if $i < k$ and, say, 0 otherwise. From 4.3(a) we see easily that $K \sim N$ and so $K \times N \sim N$. Since f maps $K \times N$ onto Alg, Alg is countable by 4.2.

A final and very often used result about countable sets is the following, which requires AC. (Its use would have simplified slightly the proof of 4.3(b); the real order was used there to avoid AC.)

Theorem 4.4^{AC}. *A countable union of countable sets is countable. (That is, if I is countable and each A_i (for $i \in I$) is countable, then $\bigcup_{i \in I} A_i$ is countable.)*

Proof. If I is finite the result is easily proved by induction on $\overline{\overline{I}}$, *without using AC* (see Problem 1). So we may as well assume $I = N$. We can further reduce to the case where each $A_i \sim N$ (by passing to $A_i' = A_i \cup N$ otherwise.) Finally, by considering the sets $\{n\} \times A_n$ and 4.2 above, we can further reduce to the case that A is disjointed. By AC we can find a function $(g_i : i \in I)$ such that for each $i \in I$, $A_i \underset{g_i}{\sim} N$. ($g_i$ can be 'picked' from a certain subset of the *set* N^{A_i}.) So $\bigcup_{i \in I} A_i \underset{h}{\sim} N \times N$, where $h(a) = (i, g(a))$ if $a \in A_i$. The theorem follows, since $N \times N \sim N$.

The need for the Axiom of Choice in proving 4.4 eluded some of the early opponents of AC, at least for a while. In fact, 4.4 is absolutely critical for much of modern analysis (dating from 1900-1920, say) and, in particular for the study of Lebesgue measure and integration, of Borel sets, and of meager sets and sets having the Baire property. Borel, Lebesgue, and Baire were all listed at one time or another among the opponents of AC!

The set R of real numbers is sometimes called the continuum, so its cardinal number is denoted by $\mathfrak{c}$ (German c). The function 'tangent' establishes a one-to-one correspondence between the open interval $(-\pi/2, \pi/2)$ and R. Using linear maps one sees that any two non-empty open intervals are equivalent. Hence one easily concludes, using the Cantor-Bernstein Theorem, that *all intervals* (open, half-open, etc.) from a to b, where $-\infty \leq a < b \leq +\infty$, have power $\mathfrak{c}$. In a similar way one easily shows

Proposition 4.5. $2^{\aleph_0} = \mathfrak{c}$. (Problem 2. Hint: Do $\leq$ and $\geq$ separately using ternary decimals in one and binary in the other.)

Hence by Cantor's Theorem (3.1),

$$\mathfrak{c} \neq \aleph_0. \tag{3}$$

Putting together (3) and 4.3(b) we have

Theorem 4.6. *There exist transcendental real numbers (i.e., reals which are not algebraic).*

Cantor gave the above proof of the existence of transcendental numbers in [F1874]. The first proof that such numbers exist was found by Liouville in 1851. Hermite (at the same time as Cantor) and Lindemann (ten years later) proved respectively that the numbers e and π are transcendental – which is of course very much more. Nevertheless, Cantor's proof was by far the shortest; it operated on a totally new principle; and it showed that in a sense nearly all real numbers are transcendental. For this reason it has attracted then and ever since much attention among mathematicians and has played a key role in convincing mathematicians of the importance of Cantor's infinite numbers.

Theorem 4.7. $\mathfrak{c} = \mathfrak{c} + \mathfrak{c} = \mathfrak{c} \cdot \mathfrak{c} = \mathfrak{c}^n = \mathfrak{c}^{\aleph_0}$.

Proof. Clearly each $\leq$ holds. We show $\mathfrak{c}^{\aleph_0} = \mathfrak{c}$, using a little idea which occurs frequently: We have $\mathfrak{c}^{\aleph_0} = (2^{\aleph_0})^{\aleph_0} = 2^{\aleph_0 \cdot \aleph_0} = 2^{\aleph_0} = \mathfrak{c}$, proving 4.7. (Note the use of the laws of Chapter 2 together with $\aleph_0 \cdot \aleph_0 = \aleph_0$.)

The 'Hilbert space' R^N and each n-dimensional space R^n are by 4.7 no bigger than R, from the standpoint of cardinal numbers! (But, from the standpoint of dimension, they *are* bigger.)

On the other hand, by Cantor's Theorem 3.1, $\mathfrak{c} < 2^{\mathfrak{c}}$, so $P(R)$ is bigger than R. We do have:

Theorem 4.8. $\mathfrak{c}^{\mathfrak{c}} = 2^{\mathfrak{c}}$ (Problem 3. Argue as above for 4.7.)

That is, the set R^R of all $f: R \to R$ and the set $P(R)$ of all subsets of R are equivalent.

It happens occasionally in algebra, and, especially, in analysis that – as a working matter – one proves existence theorems using cardinal arithmetic (as in 4.3(b)). For example, consider the notion: A is a Borel set of real numbers (of which we assume *no* knowledge!). It can easily be proved that the family of all Borel sets of reals has power $\mathfrak{c}$. Hence we can infer that there exists a set of reals which is not Borel – because $\overline{\overline{P(R)}} = 2^{\mathfrak{c}} \neq \mathfrak{c}$. For such reasons, a working mathematician needs occasionally to evaluate the cardinal number of some set he is considering. Several examples of this kind are in the problems.

Problems

1. Prove without AC: A finite union of countable sets is countable.
2. Prove 4.5
3. Prove 4.8

In problems 4-6 below show that $\overline{\overline{A}} = \aleph_0$, or that $\overline{\overline{A}} = 2^{\aleph_0}$, or else that $\overline{\overline{A}} = 2^{\mathfrak{c}}$.

4. A is the set of all continuous functions on R to R. (Hint: the rationals are dense in R.)
5. A is the set of all permutations of R. (f is a *permutation of* X if $X \underset{f}{\to} X$).
6. A is the family of all open sets of R.
7. (Fairly hard) Let $A \subseteq R$ and $\overline{\overline{A}} = \aleph_0$. (Using AC, we will later easily see from things in Chapter 7 that $\overline{\overline{R - A}} = \mathfrak{c}$.) Prove now, without using AC, that $\overline{\overline{R - A}} = \mathfrak{c}$. Hint: use Cantor's construction (proving 3.1) repeatedly in order to show that $\aleph_0 \leq \overline{\overline{R - A}}$ (which easily implies $\overline{\overline{R - A}} = \mathfrak{c}$).

Remark. By Problem 7 the set of transcendental real numbers has the power $\mathfrak{c}$.

5. ORDERS AND ORDER TYPES

§ 1. Ordered sums and products

The ordered sum of two orders (and later of two 'order types') can be defined in a natural way. If the orders $\underline{A} = (A, \leq)$ and $\underline{A}' = (A', \leq')$ are disjoint, then their *ordered sum* $\underline{A} + \underline{A}'$ is the structure $(A \cup A', \leq'')$, which is the result of 'putting $\underline{A}'$ to the right of $\underline{A}$', i.e., where, in general:

$$x \leq'' y \text{ if and only if } x \leq y \text{ or } x \leq' y \text{ or } (x \in A \text{ and } y \in A').$$

It is easy to check that $\underline{A} + \underline{A}'$ *is indeed an order*.

Passing to a corresponding infinite sum, let $\underline{I} = (I, \leq)$ be an order and suppose $(\underline{A}_i: i \in I)$ is a disjointed indexed family of orders (where each $\underline{A}_i = (A_i, \leq_i)$). Then we define the *ordered sum* $\Sigma_{i \in \underline{I}} \underline{A}_i$ to be $(\bigcup_{i \in I} A_i, \leq)$ where, in general,

$$x \leq y \text{ if and only if for the determined } i, j \text{ such that } x \in A_i \text{ and } y \in A_j, \text{ either } i < j \text{ or else } (i = j \text{ and } x \leq_i y).$$

It is as if one pictures $\underline{I}$ going from left to right and then replaces each i in I by a (squashed down) $\underline{A}_i$! A detailed but easy argument (Problem 1) shows again that $\Sigma_{i \in \underline{I}} \underline{A}_i$ *is an order*. (Note: it is convenient to allow also the case where the given $(\underline{A}_i: i \in I)$ is not disjointed. In this case we put $\Sigma_{i \in \underline{I}} A_i = \Sigma_{i \in \underline{I}} A_i'$, where $\underline{A}_i' = (A_i', \leq_i')$, $A_i' = \{i\} \times A_i$, and of course, $(i, a) \leq_i' (i, b)$ if and only if $a \leq_i b$.)

Note that $+$ is a special case of Σ. Indeed, $\underline{A} + \underline{B} = \Sigma_{i \in I} \underline{A}_i$ where $\underline{A}_0 = \underline{A}$, $\underline{A}_1 = \underline{B}$, and $\underline{I}$ is $\{0, 1\}$ with the usual order.

If $\underline{A}$ and $\underline{B}$ are orders, the *ordered product* $\underline{A} \cdot \underline{B}$ can be defined to be $\Sigma_{b \in B} \underline{A}$ (which we know is an order). It is easy to check that this is just the set $A \times B$ of "two letter words" (a, b) under the antilexographic order (i.e., dictionary order but reading words from right to left).

There is no 'perfect' notion of infinite ordered product, though some partial notions are of interest and have been studied. (See, for example, Hausdorff ([F1914] or [F1957]).

A last operation on orders (already mentioned in Chapter 1) takes the order $\underline{A} = (A, \leq)$ to the order $(A, \geq)$, *called* $\underline{A}^*$ i.e., to $\underline{A}$ 'turned around'.

A useful fact, whose proof (Problem 2) is a good exercise, is

Theorem 1.1.(a). *If* $\underline{I} = (I, \leq)$ *is a well order and for each* $i \in I$, $\underline{A}_i$ *is a well-order, then* $\Sigma_{i \in \underline{I}}\ \underline{A}_i$ *is a well-order.*

It follows at once that

(b) *If* $\underline{A}$ *and* $\underline{B}$ *are well-orders so are* $\underline{A} + \underline{B}$ *and* $\underline{A} \cdot \underline{B}$.

Problems

1. Show $\Sigma_{i \in \underline{I}}\ \underline{A}_i$ is an order if $\underline{I}$ is and each $\underline{A}_i$ is.
2. Prove 1.1(a).

Note: § 1 has not used the notion $\bar{\bar{A}}$ (or assumptions about cardinals). It could have been placed at the end of Chapter 1.

§ 2. Order types

Obviously the notion $\underline{A} \cong \underline{B}$ (among orders) is an equivalence. Hence it seems reasonable to behave as we did in § 1 of Chapter 2, where the notion $\bar{\bar{A}}$ was introduced. So we now imagine that to each order $\underline{A}$ corresponds something called $\mathrm{Tp}\,\underline{A}$ (the *order-type* of $\underline{A}$), the overall correspondence being such that for any orders $\underline{A}$ and $\underline{B}$,

2.1. Assumption. $\mathrm{Tp}\,\underline{A} = \mathrm{Tp}\,\underline{B}$ *if and only if* $\underline{A} \cong \underline{B}$. *Moreover for convenience we assume also that if the order* $\underline{A}$ *has finite cardinal n then* $\mathrm{Tp}\,\underline{A} = n$.

(As regards the second assumption, recall that by 2.2.3(i) all orders of an n-element set are isomorphic.) Order types are denoted by 'ρ', 'σ', and 'τ'. Order types of well-orders $\underline{A}$ are called *ordinals* and we write $\mathrm{Tp}\,\underline{A} = \mathrm{Ord}\,\underline{A}$ (the *ordinal* of $\underline{A}$). 'α', 'β', 'γ', 'δ', *denote ordinals.* The order types of certain familiar orders have special names. Thus: $\omega = \mathrm{Tp}(N, \leq)$; $\eta = \mathrm{Tp}(Q, \leq)$; and $\lambda = \mathrm{Tp}(R, \leq)$ (each $\leq$ being the 'usual' one).

Just as for cardinals, we can now introduce operations on order types. First we have the obvious

2.2. Proposition (a). *If* $\underline{A}$, $\underline{A}'$, $\underline{B}$, $\underline{B}'$ *are orders,* $\underline{A} \cong \underline{A}'$ *and* $\underline{B} \cong \underline{B}'$, *then* $\underline{A} + \underline{B} \cong \underline{A}' + \underline{B}'$, $\underline{A} \cdot \underline{B} \cong \underline{A}' \cdot \underline{B}'$, *and* $\underline{A}^* \cong \underline{B}^*$. (*For* +, *assume* $\underline{A}$)($\underline{B}$ *and* $\underline{A}'$)($\underline{B}'$).

Definition (b). *$\sigma + \tau$ (or $\sigma \cdot \tau$) is the unique ρ such that for some (disjoint, for +) orders $\underline{A}$, $\underline{B}$, $\mathrm{Tp}\,\underline{A} = \sigma$, $\mathrm{Tp}\,\underline{B} = \tau$ and $\rho = \mathrm{Tp}(\underline{A} + \underline{B})$ (respectively, $\underline{A} \cdot \underline{B}$). σ^* is defined similarly.*

By 1.1, for any (ordinals) α and β, $\alpha + \beta$ and $\alpha \cdot \beta$ are ordinals.

2.3. *Assume AC and Assumption 4.2.1.*
Proposition (a). *For any $(\sigma_i : i \in I)$ there exist orders $(\underline{A}_i : i \in I)$ such that $\mathrm{Tp}\,\underline{A}_i = \sigma_i$ for each $i \in I$. (Problem 1)*
Proposition (b). *If the orders $\underline{A}_i$ and $\underline{A}_i'$ are isomorphic for each $i \in I$ and $\underline{I} = (I, \leq)$ is an order, then $\Sigma_{i \in \underline{I}} \underline{A}_i \cong \Sigma_{i \in \underline{I}} \underline{A}_i'$.*
Definition (c). *If $\underline{I}$ is an order, $\Sigma_{i \in \underline{I}} \sigma_i$ is the unique ρ such that for some orders $(\underline{A}_i : i \in I)$, $\mathrm{Tp}\,\underline{A}_i = \sigma_i$ for each $i \in I$ and $\rho = \mathrm{Tp}(\Sigma_{i \in \underline{I}} \underline{A}_i)$.*

We can now easily write down names for many interesting order types. (In particular, they are useful in counterexamples in various parts of mathematics!) Here are some (distinct) *ordinals*:

$$0, 1, 2, \cdots, \omega, \omega + 1, \cdots, \omega \cdot 2\ (= \omega + \omega), \cdots, \omega \cdot 3, \cdots, \omega \cdot \omega, \cdots$$

The types ω^* (the type of the set of negative integers), ω, $\omega^* + \omega$ (the type of Z), and $\omega + \omega^*$ are all discrete (that is, are the types of discrete orders); and they are distinct (Problem 2). $(\omega^* + \omega) \cdot \eta$ is also discrete but clearly an $\underline{A}$ of this type is '*non-scattered*', i.e., has a subset of type η! Obviously, λ is the type of the open interval $(0, 1)$, and hence the closed interval $[0, 1]$ has order type $1 + \lambda + 1$.

Various laws for $+$, $\cdot$, and Σ over order types are easily proved. We list only some examples, while also mentioning some failures.

Proposition 2.4. (a) *Associative laws for + and for $\cdot$.* (b) *Commutative laws fail for + and $\cdot$ (Problem 4).* (c) *$\sigma(\tau + \tau') = \sigma\tau + \sigma\tau'$ (Problem 5).* (d) *But the law $(\alpha + \beta)\gamma = \alpha\gamma + \beta\gamma$ fails, even for ordinals (Problem 6).* (e) (a) and (c) *extend suitably to Σ, assuming AC and 4.2.1.* (f) *$(\sigma + \tau)^* = ?$ and $(\sigma \cdot \tau)^* = ?$ (Fill in the blanks for yourself!)*

If the orders $\underline{A} = (A, \leq)$ and $\underline{A}' = (A', \leq')$ have the same type σ, then clearly A and A' have the same cardinal; we define $\overline{\overline{\sigma}}$ to be this cardinal. Clearly $\overline{\overline{\sigma + \tau}} = \overline{\overline{\sigma}} + \overline{\overline{\tau}}$ and $\overline{\overline{\sigma \cdot \tau}} = \overline{\overline{\sigma}} \cdot \overline{\overline{\tau}}$.

It is clear that for any natural numbers m and n the ordinal sum $m + n$ and the cardinal sum $m + n$ coincide (fortunately!) and likewise for $m \cdot n$.

We close this chapter with a famous theorem (of Cantor!) which characterizes the order type η of the rationals.

Theorem 2.5(a). *Any two denumerable dense orders without first or last elements are isomorphic (and hence of type* η*).*

The proof employs a beautiful method which has since been used to prove a variety of facts about denumerable sets or structures. (Theorem 2.5 will not be used in later chapters, and hence can be omitted if time is very short.)

Proof. Let $\underline{A} = (A, \leq)$ and $\underline{A}' = (A', \leq')$ be such orders. Write $A = \{a_n : n \in N\}$ and $A' = \{a'_n : n \in N\}$ where $a = (a_n : n \in N)$ and a' are both one-to-one. We define the ordered couple (x_n, x'_n) for $n \in N$ by recursion as follows: If n is even, let k be the least number such that $a_k \notin \{x_i : i < n\}$, and put $x_n = a_k$. We may suppose that the $x_i (i < n)$ are in A and, say, in this 'order': $x_{q_0} < x_{q_1} < \cdots < x_{q_{n-1}}$; and also that the x'_i are in A' and in the *same* order, i.e., $x'_{q_0} < x'_{q_1} < \cdots < x'_{q_{n-1}}$. Now either (1) $x_n < x_{q_0}$, or (2) $x_n > x_{q_{n-1}}$, or (3) for a certain i, $x_{q_i} < x_n < x_{q_{i+1}}$. In each case, since $\underline{A}'$ is dense and has no first or last element, we can find an element a_k of $\underline{A}'$ belonging to the corresponding 'interval' over the x'_j's. We define x'_n to be the a'_k as above with the smallest k.

On the other hand, if n is odd we take x'_n to be the first unused a'_n and then find $x_n \in A$ by just reversing the roles of $\underline{A}$ and $\underline{A}'$ above.

Put $f = \{(x_n, x'_n) : n \in N\}$. One checks easily that f is a function, is one-to-one and that, in general, $a < b$ if and only if $f(a) <' f(b)$. Moreover, by the 'back and forth' aspect of our construction, we have insured that Dom $f = A$ (at even steps) and Range $f = A'$ (at odd steps). So $\underline{A} \cong_f \underline{A}'$ as desired.

A closely related result, also due to Cantor, is:

2.5(b). Let $\underline{A}'$ *have type* η. *Then every countable order* $(A, \leq)$ *is isomorphic to a suborder (i.e., substructure) of* $\underline{A}'$.

Sketch of proof. The case of $\underline{A}$ finite is obvious (from results in § 2 of Chapter 2). So we assume $\underline{A}$ is denumerable. As before let $A = \{a_n : n \in N\}$ and $\underline{A} = \{a'_n : n \in N\}$. We plan to define y_n for $n \in N$ recursively so that $\{(a_i, y_i) : i \in N\}$ is an isomorphism of $\underline{A}$ into $\underline{A}'$. The method is just like that in the proof of (a) but only goes from left to right (and does not distinguish between odd and even).

A characterization up to isomorphism of the order of the reals is discussed in Problem 8.

Problems

1. Prove 2.3(a).
2. Prove ω^*, ω, $\omega^* + \omega$, and $\omega + \omega^*$ are all different.
3. Fill in and prove: $1 + \omega = ?$.

4. Prove 2.4(b) (i.e., give two examples).
5. Prove 2.4(c).
6. Prove 2.4(d).
7. Show that $\eta + \eta = \eta$ but $\lambda + \lambda \neq \lambda$.
8. $B \subseteq A$ is said to be *dense in the order* $\underline{A} = (A, \leq)$ if whenever $x < y$ there exists $b \in B$ such that $x < b < y$. Prove that an order $\underline{A}$ is isomorphic to the real order if (we know 'only if' from Chapter 3) $\underline{A}$ has no first or last element, there is a countable set B dense in $\underline{A}$, and, in $\underline{A}$, every non-empty subset bounded above has a least upper bound.

6. AXIOMATIC SET THEORY

At this point (which is not determined but just seems a convenient one) we shall start to work in *axiomatic* set theory. The reader should imagine (for a short while) that we are just poised to begin Chapter 1! We still take the underlying logic to be intuitive (until Chapter 9). But we shall imagine that we are working or writing in a totally symbolic language. (*In practice,* we will still write proofs in *pure* English. However, each assertion of 'mathematical English' obviously can be translated into a purely symbolic assertion).

§ 1. A formalized language

The symbols of our language are as follows: The only non-logical symbol is $\in$. The constant logical symbols are $\sim$, $\wedge$, $\vee$, $\rightarrow$, $\leftrightarrow$, $\forall$, $\exists$, $\exists!$, $=$, and ι. (Recall that $\iota x \mathcal{P} x$ is read: the unique x such that $\mathcal{P} x$}. There are also parentheses '(' and ')', about which we will not be careful. There are also the ordinary variables $x, y, \cdots, A, B, \cdots, x', y'$, etc., which are thought of as ranging over sets. The only other symbols (used rarely) are predicate variables (of one place, two places, etc.) $\mathcal{P}$, $\mathcal{Q}$, $\mathcal{P}'$, $\cdots$ and operator variables $\mathcal{A}$, $\mathcal{B}$, $\mathcal{A}'$, $\cdots$.

We assume for now it is clear from the discussion of § 1 of Chapter 1 which expressions of this language are asserting or naming. (In particular if $\mathcal{P}$ has one place and $\mathcal{A}$ has two, then $\mathcal{P} x$ is an asserting expression while $\mathcal{A}_{uv}$ is a naming expression.)

Even in intuitive logic it is desirable to clarify our usage of '$\iota x \mathcal{P} x$' or 'the unique x such that $\mathcal{P} x$'. Trouble arises in connection with the so-called improper descriptions, i.e., descriptions $\iota x \mathcal{P} x$ for which it is not the case that there is exactly one x such that $\mathcal{P} x$. Usually in actually doing mathematics we just avoid saying $\iota x \cdots$ when it is improper. But a principle like the Separation Axiom

$$\exists B \forall x (x \in B \leftrightarrow x \in A \wedge \mathcal{P} x)$$

is most conveniently taken to remain valid when we substitute for $\mathcal{P}x$ *any* asserting expression, even if that asserting expression has inside it an improper description. For something to be valid we must certainly be able to understand it. So we must give a meaning to $\iota x \cdots$ in every case. The method (one of many) we use to do this goes back to Frege! It is very simple. We just take $\iota x \mathcal{Q} x$ to mean or denote the empty set when there is not exactly one x such that $\mathcal{Q}x$. (We need to use *some definable thing,* like $\emptyset$, though not especially the empty set.) So we adopt the

Empty Set Axiom. *There is exactly one x such that, for any y, $y \notin x$.*

Our treatment of 'ι' now reduces completely to adopting the following axiom or convention (usually considered as a logical axiom rather than one of set theory).

1.1. Convention. $y = \iota x \mathcal{P}x \leftrightarrow [\exists!\, x\mathcal{P}x \wedge \mathcal{P}y] \vee [\sim \exists!\, x\mathcal{P}x \wedge \forall z(z \notin y)]$.

Note: We nevertheless continue to follow the common practice of writing "$\iota x \mathcal{P} x$ exists" to mean: "$\exists!\, x\mathcal{P}x$".
Incidentally, and just by convention, the statement $\exists x(x = x)$ ('something exists') is considered to be logically valid.

§ 2. The axioms of set theory

The axiom system, *ZFC*, we shall consider is probably the most popular one. (The two other best-known axiom systems, *NB* and *M*, are discussed briefly in Chapter 10, § 3.) We list the axioms of *ZFC* below. (Some of these have already been stated in Chapter 1.)

Empty Set Axiom: (Above in § 1.) (Also the Convention 1.1.)

Extensionality Axiom: *If, for any x, $x \in A$ if and only if $x \in B$, then $A = B$.*

(We now adopt Definition 1.1.1, so we have the notations $\{x: \mathcal{P}x\}$ and $\{\mathcal{a}_x: \mathcal{P}x\}$.)

Separation Axiom: $\{x: x \in B \wedge \mathcal{P}x\}$ *exists.*

Doubleton Axiom: $\{t: t = x \text{ or } t = y\}$ *exists.*

Union Axiom: $\{x:$ *for some A*, $x \in A$ *and* $A \in W\}$ *exists. (That is,* $\bigcup_{A \in W} A$ *exists.)*

Power Set Axiom: $\{B:$ *for every* $x \in B, x \in A\}$ *exists. (That is,* $P(A)$ *exists.)*

The axioms listed so far constitute the theory or axiom system denoted by Z_0.

Replacement Axiom: $\{\mathcal{a}_x: x \in A\}$ *exists.*

The axioms of Infinity and Choice are taken just as before (see § 5 of Chapter 2 and § 6 of Chapter 1.) The last axiom of *ZFC*, the Axiom of Regularity, will not be used until Chapter 8; and is stated there.

Thus the theory *ZFC* has been described. The theory *Z* consists of Z_0 plus Infinity and Regularity. ('*Z*' stands, of course, for Zermelo to whom the whole axiom system is principally due). *ZF* is *Z* plus *F*, meaning the Axiom of Replacement. ('*F*' stands for Fraenkel who saw the need of adding this axiom – see the Introduction.) *ZFC* is *ZF* plus Choice, '*C*' standing for Choice. (The author has no responsibility for this 'style'.) '*ZC*', 'Z_0F', etc., should now be clear.

In fact, a lot of set theory uses only Z_0, or sometimes Z_0F.

Our axioms for *ZFC* contain some redundancies. (This is convenient when various 'subtheories' as above are considered.) For example, the Empty Set Axiom follows from Separation. Indeed by logic there is some set A, and then $\{x: x \in A \text{ and } x \neq x\}$ exists by Separation, and is empty. A more interesting example is

Proposition 2.1. *One can drop from Z_0F both Separation and Doubleton and still derive them (see Problem 1 for hints).*

The Axioms of Separation, Doubleton, Union, Power Set, and Replacement are all direct instances of the contradictory 'axiom'

(1) $\{x: \mathcal{P}x\}$ exists (in general).

This may be why some of them were never mentioned in the intuitive set theory of Chapters 1-5. However, the Axiom of Infinity appears to be such an instance in some of its (equivalent) forms (cf. 2.5.4) – like 'the set of all natural numbers exists' but not others – like 'there is a Peano structure'. So this whole discussion of instances of (1) must be taken with a grain of salt!

Problem

1. Prove 2.1. Hints: Get a two element set using $P(P(\emptyset))$; then get $\{x, y\}$ using Replacement. To get $U = \{x \in B: \mathcal{P}x\}$ use the Empty Set Axiom if U should be empty; otherwise get U as $\{\mathcal{A}_x: x \in B\}$ for a suitable $\mathcal{A}$.

§ 3. On Chapter 1

We now adopt Z_0F and argue within this theory. We will now easily show (in 3.2 below) that the existence of each case of $\{x: \ldots\}$, which was simply assumed without mention as it came up in Chapter 1, can be proved in Z_0F. This will establish the entire Chapter 1 in Z_0F, as every other step taken there was by ordinary logic or by various axioms of Z_0F already mentioned in Chapter 1. As we will then see (in § 4), the other chapters 2-5 are really axiomatic as they stand. Thus in a few pages we will have completely converted over to an axiomatic basis!

A useful though trivial theorem (of Z_0F) is

Proposition 3.1. *$\{x: \mathcal{P}x\}$ exists if there is a 'big enough B', i.e., a B such that $x \in B$ whenever $\mathcal{P}x$.*

Proof. Take $\{x \in B: \mathcal{P}x\}$.

In an order slightly different from that of Chapter 1, we now prove the existence of the following sets:

Theorem 3.2. *We establish the existence of the following sets: First, directly from axioms: $\emptyset$; $\{x \in A: \mathcal{P}x\}$; $\{\mathcal{A}_x: x \in A\}$; $\bigcup(W)$; $\mathcal{P}(A)$; $\{x, y\}$. Hence easily $\{x\}$, $\bigcup_{i \in I}\mathcal{A}_i$, (x, y), and $(\mathcal{A}_x: x \in X)$. Next: $A \cap B$ and $A - B$ (using Separation); $A \cup B = \bigcup(\{A, B\})$; $A \ominus B = (A - B) \cup (B - A)$; $\bigcap_{i \in I}\mathcal{A}_i$ if $I \neq \emptyset$ (by Separation). Now $A \times B$ and* Dom R *(see Problems 2 and 3). Next, easily,* Range R, $R[A]$. *Then R/S (Problem 4),* and $f \restriction A$. *Finally, A^B and $\Pi_{i \in I}\mathcal{A}_i$ (Problem 5).*

By 3.2, the entire Chapter 1 is established in Z_0F. Moreover, the same applies to § 1 of Chapter 5. (See the Aside at the end of that section.)

Much of Chapter 1 can be done in Z_0. We shall not bother with this question now, but return to it in Chapter 8.

Problems

Following the order listed in 3.2, prove in Z_0F the existence of:
1. $(\mathcal{A}_x: x \in X)$.
2. $A \times B$. (Hint: Use for once the actual form of (x, y).)
3. Dom R.
4. R/S.
5. A^B and then $\Pi_{i \in I}\mathcal{A}_i$.

§ 4. On Chapters 2-5

From our present point of view, Chapter 2 seems clearly to begin by passing to a new language having the non-logical (operation) symbol $\bar{\bar{\ }}$ (as in $\bar{\bar{A}}$) in addition to $\in$; and then adding to Z_0F a new axiom, namely:

(2.1.1) For any A, B, $\bar{\bar{A}} = \bar{\bar{B}}$ if and only if $A \sim B$.

In fact, we also adopted *sub rosa* some other ('hidden') axioms, as we will now see. Z_0F has two axioms (Separation and Replacement) involving predicate or operator variables. As an example, take Separation: $\exists B \forall x(x \in B \leftrightarrow x \in A \wedge \mathcal{P}x)$ (involving the predicate variable '$\mathcal{P}$'). We allow as a matter of logic that any assertion

$$\exists B \forall x\, (x \in B \leftrightarrow x \in A \wedge \dots)$$

is also valid, where ⋯ is any asserting expression. But in our new language, the expression ⋯ may involve '$\overline{\overline{}}$' as well as 'ϵ'. These new instances of Separation (and Replacement) are just the hidden new axioms. This new axiom system will be called $Z_0F^{=}$.

With this understanding all of Chapters 2, 3, 4, and § 1 of Chapter 5 can now be considered to take place *verbatim* in $Z_0F^{=}$. (In both the old, intuitive and the new, formal versions, extra axioms like AC are sometimes assumed.) Finally in Chapter 5, § 2, a third non-logical symbol 'Tp' is added, in a completely analogous way; the new axiom 5.2.1 ($\mathrm{Tp}\underline{A} = \mathrm{Tp}\underline{B} \leftrightarrow \underline{A} \cong \underline{B}$) is adopted, and, of course, Separation and Replacement are again enriched by new, hidden instances. Now, again, Chapter 5, § 2 can be read verbatim in the system $ZF_0^{=,\mathrm{Tp}}$. (Again, much of Chapters 2-5 can in fact be done without Replacement.)

In the next chapter we shall continue developing Cantor's set theory, studying especially well-orders. We will now work axiomatically as we go. We argue within Z_0F or sometimes Z_0FC, which only involve ϵ. Thus, at first, Chapters 2-5 (except § 1 of Chapter 5) will not be available. But just one feature of our study in Chapter 7 is that (in § 2 and § 4) we shall be able to *define*, using only ϵ, a notion '$\overline{\overline{A}}$' such that:

(2.1.1) For all A, B, $\overline{\overline{A}} = \overline{\overline{B}}$ if and only if $A \sim B$.

Similarly, we will define 'Tp $\underline{A}$' so that the corresponding condition (5.2.1) holds. Since $\overline{\overline{A}}$ and Tp $\underline{A}$ have been defined in terms of ϵ, the 'hidden' axioms are already provable. Thus at this point (end of § 4, Chapter 7) all of Chapters 1-5 will be available to us, and we will be back on a straight-line course.

7. WELL-ORDERINGS, ORDINALS AND CARDINALS

In this chapter we always work in Z_0F unless otherwise indicated. Thus Chapter 1 and § 1 of Chapter 5 are available as we begin.

§ 1. Well-orders

Some simple facts and terminology about well-orders were already given in and just before 1.8.4. Here are some more: In a well-order $\underline{A}$, every element x is clearly of just one of these three kinds: x is the first element; x is a *successor element* – i.e., x has an immediate predecessor; or x is a *limit element* – i.e., x has a predecessor but no immediate predecessor. The structure $(\emptyset, \emptyset)$ is a well-order. For any u, the structure $(\{u\}, \{(u, u)\})$ is a well-order (the only order for $\{u\}$). Let $\underline{A} = (A, \leq)$ be any well-order and $u \notin A$. Let B be the ordered sum (§ 1, Chapter 5) of $\underline{A}$ and the order with universe $\{u\}$. Then $\underline{B}$ is a well-order (by 5.1.1(b)), and clearly $\underline{A} = \underline{B}_u$. (Recall that '$\underline{B}_u$' denotes the substructure of $\underline{B}$ having universe *Pred* u.) The fact that

(0) every well-order $\underline{A}$ is of the form $\underline{B}_u$ for some well-order $\underline{B}$ and some u

will be technically convenient below.

The next result, 1.1(a), and its corollaries form the first of three 'fundamental theorems' about well-orders. The others are 1.2 (the Comparison Theorem) and 1.5 (Recursion).

Theorem 1.1(a). *Suppose $\underline{A} = (A, \leq)$ is a well-order and $f: A \to A$ is increasing or, what is the same, order preserving (i.e., $f(x) < f(y)$ whenever $x < y$). Then for every $x \in A$, $f(x) \geq x$.*

Proof. (Problem 1). Hint: If, for some a, $f(a) < a$, consider the first such a.

On the other hand, consider (outside our development, right now) the non-well-order $(Z, \leq)$ and the function f having $f(x) = x - 1$ for $x \in Z$. f is increasing but $f(x) < x$ even for *all* x.

Corollary 1.1(b). *No well-order is isomorphic to (even a substructure of) a proper initial segment of itself.*

(Everyone 'memorizes' the statement without the parenthetical part. Occasionally the full statement is useful.)

Proof. Suppose $\underline{A}$ is a well-order and $\underline{A} \underset{f}{\cong} \underline{B}$ where $\underline{B}$ is a substructure of $\underline{A}_a$ for a certain a. Then obviously $f(a) < a$. But $f: A \to A$ is increasing, so by (a), $f(a) \geq a$, a contradiction.

An order (or indeed *any* structure) $\underline{A} = (A, \cdots)$ is called *rigid* if it has no automorphisms except Id_A. (An *automorphism* of $\underline{A}$ is an $(\underline{A}, \underline{A})$-isomorphism.)

Corollary 1.1(c). *Any well-order $\underline{A} = (A, \leq)$ is rigid.*

Proof. (Problem 2.) Hint: Suppose $\underline{A} \underset{f}{\cong} \underline{A}$. Apply (a) both to f and to f^{-1}.

(b) and (c) imply at once:

Corollary 1.1(d). *For any well-orders $\underline{A}$ and $\underline{B}$, at most one of the following holds: (i) $\underline{A}$ is isomorphic to a proper initial segment of $\underline{B}$; (ii) $\underline{B}$ is isomorphic to a proper initial segment of $\underline{A}$; (iii) $\underline{A} \cong \underline{B}$. Moreover, in each of (i)-(iii) the corresponding map is unique.*

We now establish the very important fact that *at least one* of (i)-(iii) always holds:

Theorem 1.2. (Comparison Theorem for Well-orders.) *Of any well-orders $\underline{A}$ and $\underline{B}$, one is isomorphic to an initial segment of the other.*

Proof. Let H be the set of all f such that f is an isomorphism between an initial segment of $\underline{A}$ and an initial segment of $\underline{B}$. By 1.1(d) one easily sees (Problem 3) that

(1) H is a chain (i.e., if $f, g \in H$ then $f \subseteq g$ or $g \subseteq f$).

Using (1) one can easily show (Problem 4) that

(2) $\bigcup_{f \in H} f$ belongs to H.

Call it h. (Thus h is the $\subseteq$-largest member of H.) If either initial segment Dom h or Range h is improper we are through. Otherwise, $\underline{A}_a \cong_h \underline{B}_b$ for certain a, b. Now it is easy to check (Problem 5) that

(3) $h \cup \{(a, b)\}$ belongs to H.

Since h was the largest, this makes $h \cup \{(a, b)\} \subseteq h$, so $a \in$ Dom h = Pred a, which is absurd.

The following corollary of 1.2 and 1.1 plays a key role in the easy proof of the Cantor-Bernstein Theorem using AC (see 5.2 below).

Corollary 1.3. *Suppose $\underline{B}$ is any substructure of the well-order $\underline{A}$. Then $\underline{B}$ is isomorphic to an initial segment of $\underline{A}$.*

Proof. $\underline{B}$ is a well-order (cf. § 8 of Chapter 1). Suppose 1.3 fails. Then by 1.2 (Comparison) $\underline{A}$ is isomorphic to a proper initial segment of $\underline{B}$ and hence (clearly) to a substructure of a proper initial segment of itself. This contradicts the full version of Corollary 1.1(b) above.

We now discuss induction for a well-order. Consider two assertions about an order $\underline{A} = (A, \leq)$.

(a) If B is any non-empty subset of A, then it has a minimal element.
(b) If C is any subset of A, and, for each $a \in A$, if $x \in C$ for all $x < a$, then $a \in C$, then for all $a \in A$, $a \in C$.

(a) just says '$\underline{A}$ is a well-order'. (b) is (naturally) called the induction principle for $\underline{A}$. (The course-of-values form of induction is henceforth the primary one.) The very easy analysis (for the case of the natural numbers) made in the proof of 2.2.4(c) can be applied here essentially verbatim to show that (b) is equivalent to (a) and is indeed almost literally the contrapositive of (a). Thus we have proved

Proposition 1.4. *Every well-order satisfies the induction principle (by the very definition of 'well-order').*

(It makes no difference in 1.4 and (b) if we deal with a property $\mathcal{P}$ instead of a set C, and with showing that $\mathcal{P}a$ holds for all $a \in A$ by "$\underline{A}$-induction". One has only to take $C = \{x \in A : \mathcal{P}x\}$ and the new form involving $\mathcal{P}$ reduces to (b).)

There is an analogue for any well-order $\underline{A}$ of the other kind of induction. It follows so easily from 1.4 that it need not be stated as a theorem. Suppose we want to show that for all $a \in A$, $\mathcal{P}a$. Then it is enough to show:

(i) The first element of A (if $A \neq \emptyset$) has $\mathcal{P}$.
(ii) For any $a \in A$, if $\mathcal{P}a$ and b is the immediate successor of a then $\mathcal{P}b$.
(iii) For any limit element $a \in A$, if $\mathcal{P}x$ for all $x < a$, then $\mathcal{P}a$.

Indeed, (i)-(iii) clearly imply that $\mathcal{P}a$ whenever $\mathcal{P}x$ for all $x < a$. So 1.4 applies.

We now obtain a theorem, 1.5, on recursion over a fixed well-order $\underline{A}$. The arguments are very close to those in § 4 of Chapter 2. A minor difference is that we are now using the course-of-values form. Thus at 'stage' a (where $a \in A$) we want to imagine that we 'know' all $f(x)$ for $x < a$. The mathematical object having just this information content is the function $(f(x): x < a) = f \restriction \mathrm{Pred}\, a$. (The *set* $\{f(x): x < a\}$ does not tell which x went to which value). With this in mind, it is natural to state 1.5 as follows:

Theorem 1.5. (Recursion over a fixed well-order). ($\mathcal{a}$ *is given.*) *Let* $\underline{A} = (A, \leq)$ *be a well-order. Then there is exactly one f such that*

(*) *f is a function on A and for each $a \in A$, $f(a) = \mathcal{a}_{f \restriction \mathrm{Pred}\, a}$.*

Proof. Let us abbreviate (*) by: *f works for* $\underline{A}$. By the remark (0), it is enough to fix an arbitrary well-order $\underline{B} = (B, \leq)$ and show that

(4) For each $b \in B$, there is exactly one f which works for $\underline{B}_b$.

Clearly,

(5) If $b \in B$, $c < b$, and f works for $\underline{B}_b$ then $f \restriction \mathrm{Pred}\, c$ works for $\underline{B}_c$.

We show next that

(6) If f and g both work for $\underline{B}_b$ then $f = g$. (Problem 6. Show $f(x) = g(x)$ for all $x < b$ by induction on 'x'.)

Now we can show by induction on 'b' that (4) holds. (Problem 7. Use (5) and (6).) Thus 1.5 is proved.

In Chapter 2, § 2, from recursion for any fixed W_k (2.4.1(c)) we easily inferred a kind of 'global' recursion over *all* n (2.4.1(b) (c)). In a somewhat similar way we now infer from 1.5 a 'global' form of recursion over (the class of) all well-orders $\underline{A}$ (each $\underline{A}$ acting as a 'point').

Corollary 1.6. (a) *We can introduce a notion $\mathcal{O}(\underline{A})$ (defined for all well-orders $\underline{A}$) by saying: "$\mathcal{O}$ is defined recursively by requiring that:*

(7) $$\mathcal{O}(\underline{A}) = \mathcal{a}_{\{\mathcal{O}(\underline{A}_a):\, a \in A\}} \textit{ for each well-order } \underline{A} = (A, \leq)."$$

Indeed: (*i*) *We can define $\mathcal{O}$ explicitly so that* (7) *holds, by taking $\mathcal{O}(\underline{A})$ = the unique z such that for some f, f works for $\underline{A}$ and $z = \mathcal{a}_{\{f(a):\, a \in A\}}$.*
(*ii*) *If* (7) *holds for $\mathcal{O}$ and also for $\mathcal{O}'$, then for all well-orders $\underline{A}$, $\mathcal{O}(\underline{A}) = \mathcal{O}'(\underline{A})$.*

1.6. (b) *Let Θ be defined recursively by* (7). If $\underline{A} \cong \underline{A}'$ *(both being well-orders) then* $\Theta(\underline{A}) = \Theta(\underline{A}')$.

Proof. Consider (i). By 1.5, for any well-order $\underline{A}$, there is exactly one f and exactly one z as in the proposed definition of $\Theta(\underline{A})$. The fact that (7) holds is now easily proved using (5) and (6) above (see Problem 8). Thus (i) is proved. In (ii), by (0), it is enough to fix $\underline{B}$ and show for all $b \in B$, that $\underline{\Theta}(\underline{B}_b) = \underline{\Theta}'(\underline{B}_b)$. This can easily be done by induction on 'b' (or without induction using 1,5). The proof of 1.6(b) is instructive although not difficult (see Problem 9).

Problems

1. Prove 1.1(a).
2. Prove 1.1(c).
3. Prove (1).
4. Prove (2).
5. Prove (3).
6. Prove (6).
7. Prove (4) (given (5) and (6)).
8. Prove 1.6(a)(i). No induction is needed.
9. Prove 1.6(b).

§2. Von Neumann ordinals

On the way to introducing $\overline{\overline{\underline{A}}}$ and Tp$\underline{A}$ (as promised at the end of Chapter 6), we first introduce by definition a notion Ord $\underline{A}$ (the ordinal of $\underline{A}$), for any well-order $\underline{A}$, in such a way that we can show

(2.1) for any well-orders $\underline{A}$ and $\underline{B}$, Ord $\underline{A}$ = Ord $\underline{B}$ if and only if $\underline{A} \cong \underline{B}$.

The first person to succeed in defining Ord $\underline{A}$ adequately (i.e., in such a way that (2.1) holds) was von Neumann (see [F1923-1928]). His ordinals are exactly those we shall now introduce. (But there is nothing 'holy' about this particular notion of ordinal. Other adequate definitions can be given by different or even widely different methods. Any one definition can be regarded as 'ad hoc'!).

Roughly speaking, we plan to take the von Neumann ordinals to be the sets

$$\emptyset, \{\emptyset\}, \{\emptyset, \{\emptyset\}\}, \{\emptyset, \{\emptyset\}, \{\emptyset, \{\emptyset\}\}\}, \ldots,$$

where each is the set of all preceding ones. We now begin the precise discussion.

Definition 2.2. We define Ord $\underline{A}$, for all well-orders $\underline{A}$, by recursion, by requiring for any well-order $\underline{A} = (A, \leq)$:

$$\text{Ord } \underline{A} = \{\text{Ord } \underline{A}_a : a \in A\}.$$

This is the recursion of Corollary 1.6 with $\mathcal{A}_x = x$! (Incidentally, certain other $\mathcal{A}$ can be used here, giving other adequate notions of Ord.)

We have half of 2.1 at once, by 1.6(b):

Corollary 2.3. *If $\underline{A}$ and $\underline{B}$ are well-orders and $\underline{A} \cong \underline{B}$, then* Ord $\underline{A}$ = Ord $\underline{B}$.

Of course, as before, 'ordinal' means of the form Ord $\underline{A}$ ($\underline{A}$ well-ordered), and 'α', 'β', 'γ', 'δ' range over ordinals.

Proposition 2.4. $\alpha \notin \alpha$.

Proof. As usual, it is enough to fix an arbitrary well-order $\underline{B} = (B, \leq)$ and show that for all $b \in B$ Ord $(\underline{B}_b) \notin$ Ord $(\underline{B}_b)$. We show the latter by induction on 'b'. Assume it is true of all $x < b$. If Ord $\underline{B}_b$ belonged to itself then by Definition 2.2 it would be Ord $\underline{B}_x$ for some $x < b$, which is assumed not to belong to itself!

Theorem 2.5. *If* Ord $\underline{A}$ = Ord $\underline{B}$ *($\underline{A}$, $\underline{B}$ well-orders) then $\underline{A} \cong \underline{B}$. (Hence, by 2.3, the adequacy condition 2.1 holds.)*

Proof. By the Comparison Theorem (1.2), we would otherwise have, *say*, $\underline{A} \cong \underline{B}_b$ for some b. Then, by 2.3, Ord $\underline{A}$ = Ord $\underline{B}_b$. But by definition, Ord $\underline{B}_b \in$ Ord B. Thus Ord $\underline{A} \in$ Ord $\underline{A}$, contrary to 2.4.

The main facts about ordinals (2.6-2.11 below) are direct corollaries (via 2.1) of theorems about well-orders (in § 1). The von Neumann ordinals just defined have various special properties *not* guaranteed by (the adequacy condition) 2.1 alone. (For example, 'every member of an ordinal is an ordinal'.) These extra properties can be quite convenient; but perhaps sometimes these 'accidental' facts can even be heuristically detrimental. Consequently, *we shall make a point of proving the results* 2.6-2.11 *using only the fact that* Ord A *has been defined in some way so that the adequacy condition* (2.1) *holds.* Then at the end of this section, in 2.12 and 2.13, we discuss briefly the special properties of the von Neumann ordinals.

Definition 2.6. We say that $\alpha < \beta$ if for some well-orders $\underline{A}$ and $\underline{B}$, Ord $\underline{A}$ = α, Ord $\underline{B}$ = β and $\underline{A}$ is a proper initial segment of $\underline{B}$.

Theorem 2.7. *< 'strictly orders' the ordinals. That is:*

(a) $\alpha \nless \alpha$;
(b) *If* $\alpha < \beta$ *and* $\beta < \gamma$ *then* $\alpha < \gamma$;
(c) $\alpha \leq \beta$ *or* $\beta \leq \alpha$.

Proof. These follow at once from known facts about well-orders, namely, 1.1(b) for (a), 1.2 for (c).

Let us agree to denote by 0, 1, 2 the first three ordinals.

For any α, put $W(\alpha) = \{\beta: \beta < \alpha\}$. If $\alpha = \text{Ord}\,\underline{A}$ then $W(\alpha) = \{\text{Ord}\,\underline{A}_a: a \in A\}$. Thus, by the Replacement Axiom:

Theorem 2.8(a). *$W(\alpha)$ exists.*

Write $\underline{W}(\alpha) = (W(\alpha), \leq_{W(\alpha)})$, where $\leq_{W(\alpha)}$ is $\leq$ cut down to $W(\alpha)$.

Theorem 2.8(b). *$\underline{W}(\alpha)$ is a well-order and* Ord $\underline{W}(\alpha) = \alpha$. *Indeed, if* Ord $\underline{A} = \alpha$, *then $\underline{A} \underset{f}{\cong} \underline{W}(\alpha)$, where $f(a) = \text{Ord}\,\underline{A}_a$ for each a in $\underline{A}$.*

Proof. The first part follows from the 'indeed' part. One can see in his head that $\underline{A} \underset{f}{\cong} W(\alpha)$, for the indicated f.

By 2.8(b) we have managed to define for each α a *single* well-order $\underline{A}$ of ordinal α. In general for any equivalence (like $\cong$ here) this is a much stronger situation than just having a suitable notion of 'type'.

The next proposition is completely analogous to results in § 2 of Chapter 2.

Theorem 2.9(a). *If for some α, $\mathcal{P}\alpha$, then there is a least α such that $\mathcal{P}\alpha$.*

Proof. Choose α so that $\mathcal{P}\alpha$. If α is the least, we are through. Otherwise $U = \{\beta < \alpha: \mathcal{P}\beta\}$ is not empty and so, since $\underline{W}(\alpha)$ is well-ordered, U has a least member, which is clearly as desired.

Theorem 2.9(b). (Induction over all ordinals). *Suppose that, for all α, if $\mathcal{P}\beta$ for all $\beta < \alpha$, then $\mathcal{P}\alpha$. Then for all α, $\mathcal{P}\alpha$.*

Proof. As twice before, (b) is the contrapositive of (a).

Theorem 2.10. (Recursion over all ordinals.) *We can (justifiably) introduce a notion $\mathcal{O}(\alpha)$ (over all α) by saying: $\mathcal{O}(\alpha)$ is defined recursively by requiring that*

$$\text{for each } \alpha,\ \mathcal{O}(\alpha) = \mathcal{Q}_{(\mathcal{O}\beta:\, \beta < \alpha)}.$$

Proof. First one passes to a two-part, precise version of 2.10 by imitating either 2.4.1(b), (c) or 1.6(a) (i), (ii). Since we already have recursion on each segment W_α of the ordinals, the proof of 2.10 can easily be given just imitating the proof of 2.4.1(b), (c) or 1.6(a) (i), (ii). Alternatively without repeating those ideas again one can derive 2.10 directly from 1.6(c) (i), (ii).

The next theorem is a close relative of the famous paradox of Burali-Forti. (If $\{x: \mathcal{P}x\}$ could always be formed, certainly the set of all ordinals would exist.)

Theorem 2.11 (a). *(Burali-Forti) The set of all ordinals does not exist. Hence:* (b) *Every set K of ordinals is bounded above.*

Proof. For (a), suppose A is the set of all ordinals. Then $R = \{(\alpha, \beta): \alpha < \beta\}$ is also a set, and (by 2.7 and 2.9(a)), $\underline{A} = (A, R)$ is a well-order, say of ordinal α. Thus (by 2.8(b)) $\underline{A} \cong \underline{W}_\alpha$, which is clearly a proper initial segment of $\underline{A}$ contradicting 1.1(b). This proves (a). In (b) if K is not bounded above, then the set $\bigcup_{\alpha \in K} W_\alpha$, which exists, would be the set of all ordinals, contrary to (a). Thus 2.11 is proved. (Instead one can prove (b) directly (see Problem 1); then (a) follows at once from (b).)

Henceforth we 'give in' and understand Ord A in von Neumann's way (that is as defined in 2.2). It follows at once (from Definitions 2.2 and 2.6 (of $<$)) that

Proposition 2.12. $\alpha = \{\beta: \beta < \alpha\}$.

From 2.12 one sees at once (possibly using a result or two from 2.6-2.11 about any notion of ordinal):

Corollary 2.13.

(a) $0 = \emptyset$ (!).

(b) $\beta < \alpha$ *if and only if* $\beta \in \alpha$.

(c) *Every member of an ordinal is an ordinal.*

(d) α *is a* transitive set (*a new notion*) – *that is, for any x, y, if $x \in y \in \alpha$ then $x \in \alpha$*).

Our present definition of 'x is a (von Neumann) ordinal' is quite indirect. Here is a more direct one, which is not hard to prove. (We shall not use 2.14.)

Proposition 2.14. *x is an ordinal if and only if x is a transitive set and $(x, \in_x)$ is a well-order. (See Problem 2.)*

Side Remark. (On a typical use of ordinals outside of set theory): Let X be any subset of (say) the plane. Write X' for the set of all $x \in X$ such that x is a *limit point* of X (i.e., every neighborhood of x contains points of X different from x). Clearly we can form X'' ($= (X')'$). Continuing we get the ordinary infinite sequence:

$$X \supseteq X' \supseteq X'' \supseteq X''' \supseteq \ldots \qquad (*)$$

Now let Y be the intersection of the sets (*). Now, taking Y', Y'', etc., we get:

$$X \supseteq X' \supseteq X'' \supseteq \ldots \supseteq Y \supseteq Y' \supseteq Y'' \supseteq \ldots$$

Obviously we can go on and on and on!

It appears that, having come to almost exactly this point, Cantor was 'forced' to invent well-orderings and ordinals. (And this was his first step in set theory, just *preceding* his study of '$A \sim B$'!)

Of course, now that we have ordinals, we would deal with the example above by defining a notion X_α recursively by

$$\begin{cases} X_0 = X \\ X_{\alpha+1} = X'_\alpha \\ X_\delta = \bigcap_{\alpha < \delta} X_\alpha \ (\delta \text{ a limit ordinal}). \end{cases}$$

The sets X_α are discussed further in any book on point set topology. They are related to what is called the Cantor-Bendixson Theorem.

Problems

1. Give a new, direct proof of 2.11(b), by forming $\Sigma_{\alpha \in (K, <)} \underline{W}_\alpha$ using results from § 1 of Chapter 5.
2. Prove 2.14.

§ 3. The well-ordering theorem

Using the Axiom of Choice (added to Z_0F) we now prove the famous

Theorem 3.1. (Well-ordering Principle.)^{AC} *Every set A can be well-ordered (that is, for some $\leq$, $(A, \leq)$ is a well-order). Obviously equivalent is: For any A there exists α such that $A \sim \alpha$.*

The fact 3.1 was the cornerstone of Cantor's treatment of arbitrary infinite cardinals. Thus it will allow us to answer many questions left open in Chapters 2 and 4 (later in this chapter). The Well-ordering Principle is often called Zermelo's Theorem – although Cantor first stated it and saw its truth. This is because Zermelo, in a fundamental paper [F1904], (a) stated for the first time the Axiom of Choice, and (b) derived the Well-ordering Theorem from the Axiom of Choice.

A 'very intuitive' proof of 3.1 goes as follows. Pick an element a_0 of A. If you have not exhausted A pick another, a_1. Continue this process. Eventually A is exhausted and (obviously) A is arranged in a well-order. (This argument, as it stands, appears to most people to have a lot of merit, and also to have some dubious aspects! It is a perfect example of why it is *easier*, though not necessarily *better*(!), to work in an axiomatic system, where it is clear what to accept or reject, than to work from raw intuition, where it is not. The correct proof of 3.1 given just below can be regarded as just a clarification of the 'very intuitive one', above.

Proof of 3.1. Given A we will find β so that $A \sim \beta$. By AC, we obtain F on $P(A) - \{\emptyset\}$ such that $F(U) \in U$ for each $U \in \text{Dom } F$. For purely technical reasons, pick $z \notin A$. We define $\mathcal{O}\alpha$ for all α recursively by requiring (for all α)

$$\begin{cases} \mathcal{O}\alpha = F(A - \{\mathcal{O}\beta : \beta < \alpha\}) \text{ if } A - \{\mathcal{O}\beta : \beta < \alpha\} \text{ is not empty, and} \\ \mathcal{O}\alpha = z, \text{ otherwise.} \end{cases}$$

Case 1. For some α, $\mathcal{O}\alpha = z$. Let α be in fact the least such ordinal. Let $f = (\mathcal{O}\beta : \beta < \alpha)$. Clearly $\alpha \underset{f}{\sim} A$, as desired.

Case 2. Otherwise. Let $A' = \{\mathcal{O}\alpha \in A : \alpha \text{ is an ordinal}\}$. Then, by Replacement, the set $\{\iota\alpha(\mathcal{O}\alpha = a) : a \in A'\}$ exists, and it is clearly just the set of all ordinals, contradicting 2.11. Thus 3.1 is proved.

Over many years the Well-ordering Theorem and, in particular, its consequence that *the set R of all reals can be well-ordered* have drawn the greatest amount of attention from those doubtful of the Axiom of Choice. It is a 'pure existence' theorem, not giving any way of constructing or even defining a well-ordering $\leq$ of R. Indeed, today it is known that no *defined* $\leq$ can be proved in ZFC to well-order the set R of reals. (This is a result of S. Feferman [L1965], using the method of Paul Cohen.) Nevertheless, the Well-ordering Theorem has a remarkable group of consequences all through modern mathematics, and in particular for our work on cardinals, as we shall see in § 4-7 below.

In the rest of § 3, we introduce another very useful principle, called Zorn's Lemma, and establish (in Z_0F) the equivalence of AC, the Well-ordering Principle, and Zorn's Lemma. (No use is made later of the remainder of § 3.)

First note that trivially

3.2. *The Well-ordering Principle implies AC.*

Proof. If $(A_i : i \in I)$ are non-empty sets, well-order $\bigcup_{i \in I} A_i$ (say by $\leq$). Now define a choice function f on I by putting (for each $i \in I$), $f(i) =$ the $\leq$-first member of A_i. (The technique used here and in many other proofs is sometimes described as: "Well-order everything in sight.")

Among many statements known to be equivalent to AC, Zorn's Lemma (which appears awkward at first reading) seems to be in an especially useful and transparent form, just ready for applications in algebra, topology, etc. Other related 'maximal principles' were proved earlier by Hausdorff and Kuratowski. Zorn's Lemma is most often stated in a form which is slightly different from the one below, but obviously equivalent to it.

Zorn's Lemma. *Let $\underline{A} = (A, \leq)$ be a partial order, and suppose that every subset B of A which is ordered by $\leq$ has an upper bound. Then $\underline{A}$ has a maximal member.*

Theorem 3.3. *Zorn's Lemma is equivalent to AC.*

Proof. First assume AC and let $\underline{A} = (A, \leq)$ be as in the hypothesis of Zorn's Lemma. By AC, we can obtain a choice function F for $P(A) - \{\emptyset\}$. For technical reasons, choose $z \notin A$. We define a_α by recursion over all ordinals α by putting

$$a_\alpha = \begin{cases} F(W) \text{ if the set } W = \{x \in A\text{: for all } \beta < \alpha,\ a_\beta < x\} \text{ is not empty,} \\ z \text{ otherwise.} \end{cases}$$

If $a_\alpha \neq z$ for all $\alpha < \alpha_0$, then clearly $a_\alpha < a_\beta$ whenever $\alpha < \beta < \alpha_0$. Thus, if $a_\alpha \neq z$ for *all* α, we could obtain the set of all ordinals – as in Case 2 of the proof of 3.1, above. Hence there is an α such that $a_\alpha = z$; and we take α' to be the first such. By the hypothesis of Zorn's Lemma, the set $\{a_\alpha\text{: } \alpha < \alpha'\}$ has an upper bound, though by our construction it has no 'strict' upper bound. Thus $\{a_\alpha\text{: } \alpha < \alpha'\}$ has a largest member, say, b. Clearly b is maximal, as desired. Thus AC implies Zorn's Lemma.

Now assume Zorn's Lemma and let the sets $(A_i\text{: } i \in I)$ all be non-empty. (The proof which follows is a typical argument using Zorn's Lemma.) Let K be the family of all f such that Dom $f \subseteq I$ and for each $i \in$ Dom f, $f(i) \in A_i$; and consider $(K, \subseteq_K)$. Treating each f as a set of ordered couples, it is clear that the union of a chain C of f's belonging to K also belongs to K (and hence is an upper bound for C). Hence we can apply Zorn's Lemma, obtaining a maximal member g of K. If Dom $g = I$ then g is a choice function as desired. Otherwise choose $i_0 \in I -$ Dom g, and let $a \in A_{i_0}$. Clearly $g \cup \{(i_0, a)\}$ also belongs to K, contradicting the maximality of g. Thus 3.3 is proved.

The problems below give some typical and well-known examples of the use of Zorn's Lemma in ordinary mathematics. (Some notions from modern algebra are assumed.)

Problems

Assume AC.

1. In any commutative ring A, every ideal is included in a maximal ideal. (A itself is not considered to be an ideal. "Maximal" means in the sense of $\subseteq$.)
2. In any vector space V over a field F there exists a basis B, i.e. a set $B \subseteq V$ which is independent and generates V. Hint: Obtain a maximal independent set and show it is a basis. (This is a common strategy.)
3. Let $(A, \leq)$ be any partial order. Show $\leq$ can be enlarged to an order. That is, show there exists $\leq'$ such that $(A, \leq')$ is an order and $\leq \subseteq \leq'$.

§ 4. Defining $\bar{\bar{A}}$ and Tp$\underline{A}$

We adopt AC for the rest of this chapter. We can now introduce by definition (using only $\in$) an adequate notion of '$\bar{\bar{A}}$':

Definition 4.1. We put $\bar{\bar{A}}$ = the least ordinal α such that $A \sim \alpha$. (One exists by the Well-ordering Theorem.)

The desired property (2.1.1) is immediate:

Corollary 4.2. $\overline{\overline{A}} = \overline{\overline{B}}$ *if and only if* $A \sim B$.

As remarked at the end of Chapter 6, since $\overline{\overline{A}}$ has been defined using only $\in$ we now have both the direct assumption (2.1.1) and the 'hidden' assumptions made at the beginning of Chapter 2.

An ordinal α is called an initial ordinal if α is not set-theoretically equivalent to any smaller ordinal. (Aside: Note that 'initial ordinal' makes sense in the absence of AC.) Obviously, *α is a cardinal if and only if α is an initial ordinal.* It is also clear that

(4.3) $$\overline{\overline{\kappa}} = \kappa.$$

Only one more assumption was ever made in Chapters 2-4. It can also now be proved:

Theorem 4.4. *Assumption 4.2.1 holds; that is, given $(\kappa_i : i \in I)$ there exists F on I such that for each $i \in I$, $\overline{\overline{F(i)}} = \kappa_i$. (By 4.3, take $F(i) = \kappa_i$!)*

Thus *we have now all the results of Chapters* 2-4. Here are two remarks about *finite* cardinals: Since no finite set is equivalent to a proper subset, it is clear that

4.5. (a) *Every finite ordinal is a cardinal.*

It is also obvious that

(b) If $\underline{A} = (A, \leq)$ *is a finite order then* $Ord\ \underline{A} = \overline{\overline{A}}$.

The notion '$\text{Tp}\,\underline{A}$' (for $\underline{A}$ any order) of Chapter 5 can also be subsumed, if we make the definition below of $\text{Tp}\,\underline{A}$. This definition – the first definition of 'order type' within *ZFC* (and, in fact, also avoiding Regularity) – was given by A.P. Morse in his lectures in Berkeley in the late 1940's. Let $\underline{A} = (A, \leq)$ be any order and $\overline{\overline{A}} = \kappa$. In its simplest form, Morse's definition takes $\text{Tp}\,\underline{A}$ to be

(1) the set of all orders (κ, R) isomorphic to $\underline{A}$.

(This set is easily seen to exist, if one considers $P(\kappa \times \kappa)$, etc.) Morse's $\text{Tp}\,\underline{A}$ is a relative of Frege's $\{\underline{B} : \underline{A} \cong \underline{B}\}$, but, unlike Frege's, is a *set*. It is immediate that for any orders $\underline{A}$, $\underline{B}$,

(2) $$\text{Tp}\,\underline{A} = \text{Tp}\,\underline{B} \text{ if and only if } \underline{A} \cong \underline{B}.$$

However, it would be very convenient for us if when $\underline{A}$ is a well-order, the new $\text{Tp}\,\underline{A}$ equalled the $\text{Ord}\,\underline{A}$ of § 2 above. Fortunately, this can be accomplished in the crudest way, because of the (accidental, technical) fact:

Proposition 4.6. *No ordinal is of the form* $\text{Tp}\,\underline{A}$.

Proof. It is enough to show (*) no ordinal is a non-empty set of ordered couples – as Tp$\underline{A}$ clearly is. Since every ordinal is a set of ordinals, (*) follows from: (**): No ordinal α is an ordered couple. If $\alpha > 2$, (**) is obvious as $\overline{\overline{\alpha}} > 2$. The cases $\alpha = 0, 1, 2$ are left to the reader.

We now replace (1) by the official

Definition 4.7. Let $\underline{A}$ be an order of power κ. If $\underline{A}$ is not a well-order, put Tp$\underline{A}$ = the set of all orders (κ, R) isomorphic to $\underline{A}$; and if $\underline{A}$ is a well-order, put Tp$\underline{A}$ = Ord $\underline{A}$.

From 4.6 and (2) follows:

Theorem 4.8. *For any orders $\underline{A}$, $\underline{B}$,* Tp$\underline{A}$ = Tp$\underline{B}$ *if and only if* $\underline{A} \cong \underline{B}$.

We also have our last, minor requirement made just after 5.2.1. Indeed

Proposition 4.9. *If* $\underline{A} = (A, \leq)$ *is an order and* $\overline{\overline{A}} = n$, *then* Tp$\underline{A} = n$ *(by 4.5(b) above).*

We now have (for our $\overline{\overline{A}}$ *and* Tp$\underline{A}$*) all of the results of Chapters* 2-5. (Of course Infinity must be assumed when it was there.)

Aside. Consider the following development: Chapter 1, Chapter 6, § 1-4 of Chapter 7, and then Chapter 2 except § 5. The Axiom of Infinity is never used. The arithmetic of +, ·, etc., is developed. This is simply number theory (with a set-theoretic 'style'). Thus, by taking no stand on the Axiom of Infinity, we have simultaneously developed the beginnings of *axiomatic number theory* (obtained by adding as an axiom the negation of the Axiom of Infinity) and of *axiomatic set theory* (obtained by adding the Axiom of Infinity).

Problem

1. Consider groups $\underline{G} = (G, \cdot)$. Define a notion IT($\underline{G}$) (the isomorphism type of $\underline{G}$) in such a way that for all groups $\underline{G}$, $\underline{G}'$, IT($\underline{G}$) = IT($\underline{G}'$) if and only if $\underline{G} \cong \underline{G}'$.

§ 5. Easy consequences for cardinals of the Well-ordering Principle

In § 5 and § 6 we will be able to answer many questions left unresolved in Chapters 2-5. Since cardinals are now certain special ordinals, we have to distinguish between $\kappa \leq \lambda$ (ord) – i.e., in the ordinal sense, and $\kappa \leq \lambda$ (card) – i.e., in the cardinal sense. However, this is necessary only very briefly because of the following theorem:

Theorem 5.1. *$\kappa \leq \lambda$ (card) if and only if $\kappa \leq \lambda$ (ord).*

Proof. The arrow ($\leftarrow$) is obvious, since if $\kappa \leq \lambda$ (ord) then in fact $\kappa \subseteq \lambda$.
Suppose $\kappa \leq \lambda$ (card). Then $\kappa = \overline{\overline{B}}$ for some $B \subseteq \lambda$. Now, by 7.1.3, $\underline{B}$ (the substructure of $\underline{\lambda} = (\lambda, \leq_\lambda)$ having universe B) is isomorphic to an initial segment of $\underline{\lambda}$. Thus letting $\beta = \text{Ord}\, \underline{B}$ we have $\beta \leq \lambda$ (ord). But also, $\kappa \leq \beta$ (ord), since $\overline{\overline{B}} = \overline{\overline{\beta}} = \kappa$ and κ is an initial ordinal. Hence $\kappa \leq \lambda$ (ord) as desired.

Henceforth, of course, we just write '$\kappa \leq \lambda$'.

By 5.1 the cardinals with their natural order are (except for not forming a set) a 'suborder' of the usual ordering of the ordinals. Hence we have at once 5.2 and 5.3 below (which answer questions raised just after 2.1.6(c)).

Theorem 5.2. *(Cantor-Bernstein – the short proof using AC promised in § 1 of Chapter 4). If $\kappa \leq \lambda$ and $\lambda \leq \kappa$ then $\kappa = \lambda$.*

Notice that 5.2 depended on 5.1, and the proof of 5.1 made a key use of 7.1.3 (on substructures of well-orderings) as was promised just before 7.1.3. Of course the really big thing behind both 5.2 and 5.3 is the well-ordering theorem. 5.2 has a longer proof avoiding *AC*, but 5.3 has none.

Theorem 5.3. (Comparability of cardinals.) *$\kappa \leq \lambda$ or $\lambda \leq \kappa$.*

From 5.1 we also see at once that: Every non-empty set (or even 'class') of cardinals has a least member. In particular for any κ, since there is a larger cardinal, namely 2^κ, there is an *immediate successor* of κ (among cardinals); *it is denoted by* κ^+. (In Problem 2, a set of power κ^+ is directly formed.)

We now adopt the Axiom of Infinity for the rest of Chapter 7. (Thus we are now working in *ZFC* – {*Reg*}). Below is a picture of the ordinals and cardinals (+ is the sum of order types from Chapter 5, $\cdot$ is the product there.

$\underline{0}, \underline{1}, \underline{2}, \underline{3}, \cdots, \underline{\omega}, \omega + 1, \omega + 2, \cdots, \omega \cdot 2, \omega \cdot 2 + 1, \cdots, \omega \cdot 3, \cdots, \cdots \omega \cdot \omega, \cdots$

The ones underlined are the initial ordinals (or cardinals). The first cardinal, ω^+, greater than ω is *way* off the page! The next is even farther beyond ω^+. Nevertheless the cardinals never stop occurring in the list of ordinals. A convenient notation for the infinite cardinals is as follows:

We define the cardinals $\aleph_\alpha$ for all α by recursion, putting

$$\aleph_\alpha = \text{ the first infinite cardinal greater than } \aleph_\beta \text{ for all } \beta < \alpha.$$

Such a cardinal exists because $\{\aleph_\beta : \beta < \alpha\}$ has an upper bound λ by 4.2.5(b), and hence a strict upper bound λ^+.

Of course, $\aleph_0 = \omega$ and $\aleph_{\alpha+1} = (\aleph_\alpha)^+$. It is immediate from the definition that if $\alpha < \beta$ then $\aleph_\alpha < \aleph_\beta$. The other thing we certainly intended is

Proposition 5.4. *Every infinite κ is an 'aleph' (i.e., is of the form $\aleph_\alpha$). 5.4 is 'obvious'; but finding a proof takes some thought (Problem 2).*

Aside. The notation ω_α is also used to mean $\aleph_\alpha$. This redundant notation is used in a remarkable but widely employed notational convention, to resolve the ambiguity of cardinal + and ordinal + (and likewise for $\cdot$)! Thus $\aleph_\alpha + \aleph_\beta$ means the cardinal sum, but $\omega_\alpha + \omega_\beta$ means the ordinal.

Problems

1. 5.2 and 5.3 can instead be proved first, and then used to prove 5.1. Derive 5.3 using 1.2, and 5.2 using 5.3 and 1.3. Now infer 5.1, using 1.8.2.
2. Prove 5.4.
3. Suppose $\overline{\overline{A}} = \kappa$. We can define a well-ordering of type κ^+ as follows. Let K be the set of all well-orders $\underline{B} = (B, R)$ such that $B \subseteq A$. Write $\underline{B}\ E\ \underline{B}'$ for $\underline{B} \cong \underline{B}'$. Let $W = K/_E$ = the set of all equivalence classes $\underline{B}/_E$ for $\underline{B} \in K$. Show that the definition: "$\underline{B}/_E \leq \underline{B}'/_E$ if and only if $\underline{B}$ is isomorphic to an initial segment of $\underline{B}'$" is justified; and show that $\leq$ well-orders W in the order type κ^+.
4. Show there are α such that $\omega_\alpha = \alpha$. (Hint: Take sup $(\omega, \omega_\omega, \omega_{\omega_\omega}, \cdots)$; but do it properly using recursion.)

§ 6. A harder consequence and its corollaries

We can now prove the important

Theorem 6.1. *If κ is infinite, then $\kappa \cdot \kappa = \kappa$.*

Proof. Of course, $\kappa \leq \kappa \cdot \kappa$. We prove 6.1 by *induction on cardinals.* (This is obviously justified since the cardinals form a 'suborder' of the ordinals.) Suppose $\lambda \cdot \lambda = \lambda$ for all infinite $\lambda < \kappa$. Suppose κ is infinite. Glance at the Figure below.

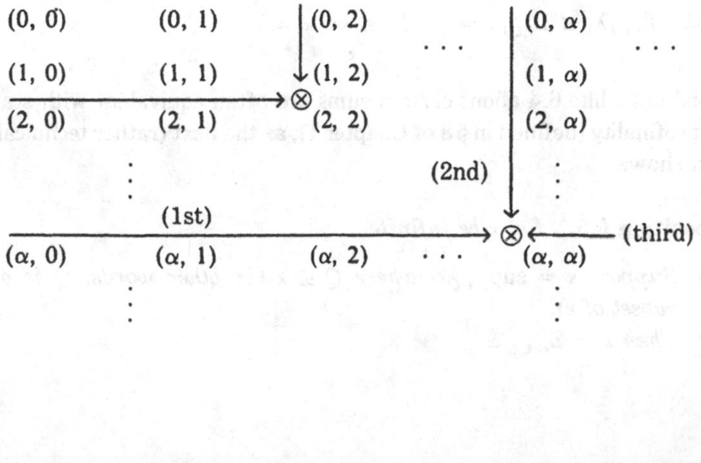

Clearly

$$\kappa \times \kappa = \bigcup_{\alpha < \kappa} [(\{\alpha\} \times \alpha) \cup (\alpha \times \{\alpha\}) \cup \{(\alpha, \alpha)\}],$$

all those unions being 'disjointed'. Hence we can define in the obvious way on $\kappa \times \kappa$ an order $\leq$ of *order type* $\Sigma_{\alpha < \kappa}(\alpha + \alpha + 1)$ (ordinal sums). (See figure.) In fact, by 5.1.1, a well-ordered sum of well-orders is a well-order, so $(\kappa \times \kappa, \leq)$ has order type δ, an *ordinal.* Any proper initial segment of $(\kappa \times \kappa, \leq)$ is clearly included in another which has cardinal $\mu = \Sigma_{\alpha < \beta}(\overline{\overline{\alpha}} + \overline{\overline{\alpha}} + 1)$ for some $\beta < \kappa$. We will show $\mu < \kappa$. Indeed, if β is finite, then clearly μ is finite (so $< \kappa$). Suppose β is infinite. Then $\overline{\overline{\alpha}} + \overline{\overline{\alpha}} + 1 \leq 3\overline{\overline{\beta}} \leq \overline{\overline{\beta}} \cdot \overline{\overline{\beta}} = \overline{\overline{\beta}}$, by the inductive hypothesis. Thus $\mu \leq \Sigma_{\alpha < \beta}\overline{\overline{\beta}} = \overline{\overline{\beta}} \cdot \overline{\overline{\beta}} = \overline{\overline{\beta}}$ (again). Thus every initial segment of $(\kappa \times \kappa, \leq)$ has cardinal $< \kappa$; so certainly $\kappa \not< \delta$. Thus $\delta \leq \kappa$ and $\kappa \cdot \kappa = \overline{\overline{\delta}} \leq \kappa$ as desired.

6.1 has a number of important corollaries.

Corollary 6.2.

(a) *If κ is infinite then $\kappa + \kappa = \kappa$.* (By 6.1, since, by 2.3.2 (g), $\kappa + \kappa \leq \kappa \cdot \kappa$).

(b) *If at least one of κ and λ is infinite, then*

$$\kappa + \lambda = \kappa \cdot \lambda = max\ (\kappa, \lambda)\ (at\ once).$$

6.1(a) was proved (in § 4 of Chapter 4) if $\kappa = \aleph_0$ or $\kappa = 2^{\aleph_0}$ (*without using AC*). We can now extend to all infinite κ (using *AC*) a little argument made there, getting:

Corollary 6.3. *If κ is infinite then* (a) $(2^\kappa)^\kappa = 2^\kappa$; *and* (b) $\kappa^\kappa = 2^\kappa$. *(See Problem 1.)*

Corollary 6.4. *If κ is infinite, $\overline{\overline{I}} \leq \kappa$, and $\lambda_i \leq \kappa$ for each $i \in I$, then* $\Sigma_{i \in I}\lambda_i \leq \kappa$. *(Any union of at most κ sets, each of power at most κ, has power at most κ.)*

Proof. $\Sigma_{i \in I}\lambda_i \leq \Sigma_{i \in I}\kappa = \kappa \cdot \overline{\overline{I}} \leq \kappa \cdot \kappa = \kappa.$

Conditions like 6.4 about certain sums are often equivalent with statements about cofinality (defined in § 8 of Chapter 1), as the next (rather technical) proposition shows.

Proposition 6.5. *Let κ be infinite.*

(a) *Suppose $\kappa = \sup_{\alpha \in Q}\alpha$ where $Q \subseteq \kappa$ (in other words, Q is a cofinal subset of κ).*
Then $\kappa = \Sigma_{\alpha \in Q}\overline{\overline{\alpha}}$

(b) *Suppose* $\Sigma_{i \in I}\lambda_i = \kappa$ *where* $\overline{\overline{I}} < \kappa$ *and, for each* $i \in I$, $\lambda_i < \kappa$. *Put* $Q = \{\lambda_i : i \in I\}$. *Then* $\kappa = \sup_{i \in I}\lambda_i$.

Proving 6.5 (see Problem 2) is not difficult, but it definitely requires care. (Cofinality is studied further in § 1 of Chapter 11.)

An infinite cardinal λ is called *regular* if λ *has no cofinal subset of power less than* λ (otherwise, λ is *singular*). By 6.5 we can also say: λ *is regular if and only if* λ *cannot be expressed as a sum of fewer than* λ *cardinals each less than* λ.

It is immediate from the definition that $\aleph_0$ is regular. Using 6.4, we also see at once the important fact:

Theorem 6.6. *If* κ *is infinite, then* κ^+ *is regular.*

The remaining infinite cardinals are the limit cardinals greater than $\aleph_0$. The first one, $\aleph_\omega$, is clearly singular, since $\aleph_\omega = \Sigma_{n \in \omega}\aleph_n$. In a similar way one sees that, in general:

(1) If α is a countable limit ordinal, then $\aleph_\alpha$ is singular (Problem 3).

The next cardinal, $\aleph_{\omega_1}$, is obviously singular. One can continue in this way very far. Thus the question arises whether there are any regular limit cardinals at all except $\aleph_0$. Regular limit cardinals are also called *weakly inaccessible* because they cannot be reached from below either as the next cardinal or as a sup of a smaller number of smaller cardinals. As early as 1914, F. Hausdorff [F1914] wrote: "[Regular limit cardinals $> \aleph_0$] are as yet not known; they must be of exorbitant size."

In Sierpinski and Tarski [L1930], a perhaps even more unreachable kind of cardinal was introduced. An infinite cardinal κ is called (plain) *inaccessible* (sometimes "strongly inaccessible") if κ is regular and $2^\lambda < \kappa$ whenever $\lambda < \kappa$. Of course, $\aleph_0$ *is inaccessible.* Obviously *an inaccessible cardinal is weakly inaccessible*, as always $\lambda^+ \leq 2^\lambda$.

Let us assume *ZFC* is consistent. We will show in 2.2 of Chapter 10 that in *ZFC one cannot prove:* (*) *there exists an inaccessible cardinal* $> \aleph_0$. (The idea of this proof goes back to Kuratowski [L1925].) Obviously, under *GCH*, 'inaccessible' and 'weakly inaccessible' are the same. Thus using the result of Gödel (discussed in the next section) that *GCH* is consistent with *ZFC* it follows at once that: *One cannot prove in ZFC the existence of* (even) *a weakly inaccessible cardinal.*

On the other hand, most people consider that the statement (*) is an intuitively reasonable axiom which might well be added to *ZFC*. The study of 'large cardinals' (like these) turns out to be related to the study of models of set theory, which lies on the borderline of set theory and logic. (We return to the subject of large cardinals in 10.2.4.)

Problems

1. Prove 6.3 ((a) and (b)).
2. Prove 6.5 ((a) and (b)).
3. Prove (1).

§ 7. The Continuum hypothesis

From § 4-6 it might appear that with the aid of the Well-ordering Theorem one can answer any reasonable question about cardinal numbers (except possibly about very large ones). In fact, almost the opposite is true! Cantor himself, having shown that $\aleph_0 < 2^{\aleph_0}$, tried to prove the following famous assertion:

7.1. The Continuum Hypothesis (*CH*). $2^{\aleph_0} = \aleph_1$. *(That is, there is no cardinal between* $\aleph_0$ *and* $2^{\aleph_0}$.*)*

The Continuum Hypothesis can be restated without using the notion of cardinal number and, indeed, in a way which could be understood in an introductory analysis course. Namely: Any set of real numbers can either be mapped one-to-one into the set of natural numbers or else can be mapped one-to-one onto the set of all real numbers.

Cantor himself was unable to prove either *CH* or its negation. It is natural to consider also:

7.2 The Generalized Continuum Hypothesis (*GCH*). *For any* α, $2^{\aleph_\alpha} = \aleph_{\alpha+1}$.

For nearly a century the Continuum Problem (of resolving *CH*) has stood as one of the most famous and most fundamental problems of mathematics. For example, the great David Hilbert worked hard at it and even claimed erroneously for a time to have proved *CH*!

In 1938, Kurt Gödel (F1940) – without answering *CH* either way – established the remarkable result: *One cannot prove in ZFC the negation of GCH (i.e., GCH is consistent with ZFC), assuming that ZFC itself is consistent.* Almost thirty years later, in 1963, Paul Cohen [1966] established the, if anything, even more remarkable result: *One cannot prove CH in ZFC if ZFC is consistent.* Each of these two results is generally considered to be among the greatest in mathematics in the twentieth century. Their assertions (ignoring the proofs) can be grasped by the reader fairly well now, but they will be clearer after the reader has studied the basic logic given in Chapter 9. The proofs (of Gödel and of Cohen) are a central topic of many books or courses in set theory at the level just above that of this book.

It is natural to suggest, say, regarding *CH*, that we should look for other intuitively obvious axioms to add to those of *ZFC* such that on this new basis, we can resolve *CH* (that is, prove *CH* or prove its negation). Technically, this possibility has not been completely ruled out. Perhaps a third or a half of set theorists today regard it as at least a live possibility. But there are technical results which are discouraging (though not damning). Indeed, certain sentences are known not be provable in *ZFC* if *ZFC* is consistent, and yet seem to nearly everyone to be clearly valid. One example is a sentence which says (within set theory) "*ZFC* is consistent" – from Gödel's most famous paper [1931]. Another example (whose validity seems a bit more dubious) is "There exists a weakly inaccessible cardinal." (This is *proved* to be an example in § 2 of Chapter 10.) Thus one might hope to resolve *CH* by adding one of these two sentences to *ZFC*. But, in fact, *neither CH nor its negation can be derived from ZFC enriched with the above new axioms* – if *ZFC* plus these axioms is consistent. This followed from the ideas in the original proofs of Gödel and Cohen.

For each α, there is a certain $\beta(\alpha)$ such that $2^{\aleph_\alpha} = \aleph_{\beta(\alpha)}$. Can we say anything at all about $\beta(\alpha)$? It is known from work of Solovay, Easton and others (using Cohen's method) that, roughly speaking, very little can be proved in *ZFC* about the operator β, or even about $\beta(0)$. However, some things can be said. There is one old group of results about $\beta(\alpha)$, which will be discussed in § 2 of Chapter 11. For example, it will be shown there that $2^{\aleph_0} \neq \aleph_\omega$!

Another, rather recent result about the operator β is discussed in Problem 1. Quite recently an important new group of results about β was initiated by J. Silver [L 1975] and carried further by Galvin, Hajnal, Baumgartner, Prikry, and very recently, Shelah. (See Levy [1979], pp. 181-186.) To give just one example, Silver showed that if $2^\lambda = \lambda^+$ for every $\lambda < \kappa = \aleph_{\omega_1}$, then also $2^\kappa = \kappa^+$.

CH and *GCH* are not in any way special for being unresolvable in *ZFC*. Rather, using Paul Cohen's methods, a whole group of modern set theorists have shown that nearly all of the famous problems which arose in set theory from Cantor's time on, e.g., the famous problems of Souslin and Kurepa (see Levy [1979]), are also unresolvable in *ZFC*, if it is consistent.

Aside. We mentioned above that Gödel [1931] proved that (for example) in *ZFC* one cannot prove "*ZFC* is consistent" if *ZFC* is consistent. The metatheory is much weaker than *ZFC*, so we cannot hope to prove in the metatheory that *ZFC* is consistent even if it is. For this reason, results like Gödel's "*ZFC* + *GCH* is consistent" (mentioned above) and several results proved in Chapter 10 are all bound to need a curious 'waiver' in their formulation – like 'if *ZFC* itself is consistent.'

Problems

1. Let κ be singular and suppose 2^λ is eventually constant below κ, that is: for some $\lambda_0 < \kappa$ and some μ we have $2^\lambda = \mu$ for all λ such that $\lambda_0 \leq \lambda < \kappa$. Show that $2^\kappa = \mu$ (also). (Hint: Write $\kappa = \Sigma_{\alpha < \kappa'} \lambda_\alpha$, where $\kappa' < \kappa$ and each $\lambda_\alpha < \kappa$. Compute 2^κ using laws from Chapter 4, § 2.) (This result is due to Bukovsky [L1965] and Hechler [L1973].)

8. THE AXIOM OF REGULARITY

§ 1. Partial universes and the axiom of regularity

We work in Z_0F. (Thus we can depend on Chapter 1 and § 1 of Chapter 5; and especially on § 1, on well-ordering, and § 2, on ordinals, from Chapter 7.)

By recursion, put

$$\begin{cases} V_0 = \emptyset \\ V_{\alpha+1} = P(V_\alpha) \text{ for any ordinal } \alpha \\ V_\delta = \bigcup_{\alpha < \delta} V_\alpha \text{ for any limit ordinal } \delta. \end{cases}$$

Sets of the form V_α will be called *partial universes.* (More or less the same things have also been called 'types', '*Stufe*' ('steps') and 'natural models'.)

Proposition 1.1

(a) *V_α is a transitive set (i.e., if $x \in y \in V_\alpha$ then $x \in V_\alpha$).*

(b) *If $\alpha < \beta$ then $V_\alpha \subseteq V_\beta$.*

(c) *V_α is supertransitive (i.e., transitive and if $x \subseteq y \in V_\alpha$ then $x \in V_\alpha$).*

(a), (b), and (c) are easily proved by induction (Problem 1). One might expect that a 'partial universe' would be as in (a) and (c).

We write Vx to mean that for some α, $x \in V_\alpha$ (that is, x belongs to some partial universe). Obviously,

1.1 (d) *The predicate V is 'supertransitive'; i.e., if $x \in y$ or $x \subseteq y$, and Vy then Vx.*

We now introduce a notion of *rank* (in a way which is available whenever one has a 'tower' such as the V_α's form). Indeed, if Vx then we put $\mathrm{Rk}\, x$, the *rank of x, equal to the least α such that* $x \in V_{\alpha+1}$ (or, what is the same, $x \subseteq V_\alpha$). The following properties of rank are easily proved (Problems 2).

Proposition 1.2

(a) *If* Vx, Vy, *and* $x \in y$, *then* $\mathrm{Rk}\,x < \mathrm{Rk}\,y$.
(b) $V_\alpha = \{x\colon Vx \text{ and } \mathrm{Rk}\,x < \alpha\}$.
(c) *For any ordinal* α, $V\alpha$ *and* $\mathrm{Rk}\,\alpha = \alpha$.

Theorem 1.3. *If, for any* $y \in x$, Vy, *then* Vx.

Proof. Let $\beta = \sup_{y \in x}(\mathrm{Rk}\,y + 1)$. Then $x \subseteq V_\beta$, so $x \in V_{\beta+1}$ and Vx.

Is it true that for every x, Vx? (That is, does every set belong to some partial universe?) The question has a clear intuitive meaning. The Axiom of Regularity is just the proposition:

Axiom of Regularity (*Reg*). *Every set belongs to some partial universe.*

(Two other names, Axiom of *Foundation* and Axiom of *Restriction*, are used almost as often.)

It is known that the Axiom of Regularity cannot be derived from the other axioms of *ZFC*, if they are consistent – see Bernays [F1954]. In Chapter 10 we will give the proof, due to von Neumann, that, on the other hand, the Axiom of Regularity is consistent with the other axioms of *ZFC*, if they are consistent. This proof actually shows that if we throw out all sets not in V, the remaining sets (those in V) satisfy *ZFC*. Because of this one can reasonably adopt *Reg* without professing belief in it; he simply decides to focus his attention on the sets in V (and call only them 'sets'). Regarding the question whether the Axiom of Regularity is *intuitively true*, alas, some qualified people say 'yes' and other qualified people say 'no'! In any case, the rather wide 'acceptance' (in some sense) of the Axiom of Regularity, especially in the last two decades or so, is indicated by the fact that just in that period '*Reg*' has pretty generally been taken to be part of *ZFC*! Nevertheless, the Axiom of Regularity has been very little used by mathematicians. Even the mathematics of Cantor's set theory, presented here up to now, has never required Regularity.

We will next study some (more or less) equivalent forms of the Axiom of Regularity. Consider these possible axioms:

(1) Every non-empty set A has an $\in$-minimal element x (that is, for any $z \in A$, $z \notin x$).

Axiom schema (2) below says the same for any class or predicate $\mathcal{P}$, that is:

(2) If for some x, $\mathcal{P}x$ then there is an $\in$-minimal x such that $\mathcal{P}x$ (i.e., no $y \in x$ has $\mathcal{P}y$).

Also consider (3) below (only discussed when *Infinity* is assumed):

(3) There is no ordinary infinite $\in$-descending sequence (that is, one cannot have f on ω such that for each n, $f(n+1) \in f(n)$).

The next theorem might be summarized by saying that, over Z_0F, *Reg* and (2) (a schema) are 'equivalent'.

Theorem 1.4.
(a) *If Reg then* (2).
(b) *Conversely, Reg follows from a certain* $\in$*-instance of schema* (2), *namely: If, for some* x, *not* Vx, *then there is an* $\in$*-minimal* x *such that not* Vx.

Proof. For (a), take an element in $\mathcal{P}$ having the least possible rank. It is clearly the desired minimal element. For (b), assume the given instance of (2), and also that *Reg* fails. Then some x is not in V and so by assumption some minimal x not in V exists. But then every member of x is in V, so by 1.3 above, x is in V, a contradiction.

Next, *assuming the Axiom of Infinity,* we will show in 1.6 below that *Reg* and (1) are equivalent. For this purpose we need first another theorem which is of interest in itself. *Tcx* (the *transitive closure* of x) is defined to be the $\subseteq$-smallest transitive set including x. An arbitrary intersection of transitive sets is clearly transitive, so '*Tcx* exists' is the same as 'x is a subset of *some* transitive set.'

Theorem 1.5. *Infinity implies that* Tcx *exists (for every* x.) *(Obviously, so does Reg.)*

Proof. For $\mathrm{Tc}x$ take $x \cup \bigcup(x) \cup \bigcup(\bigcup(x)) \cup \cdots$ (which exists, by Replacement). (For a detailed proof see Problem 3).

Theorem 1.6. *Assume Infinity. Then Reg and* (1) *are equivalent (and so* (2) *also).*

Proof. We know that *Reg* is 'equivalent' to (2), of which (1) is an instance. So for 1.6 it is enough to derive (2) from (1). Assume (1) and assume $\mathcal{P}x$. x is itself a minimal set having the property P unless $A = \{y \in X: \mathcal{P}y\}$ is not empty. In that case, by (1), we can take z to be an $\in$-minimal member of the set $\{u \in \mathrm{Tc}A: \mathcal{P}u\}$. Since $\mathrm{Tc}A$ is transitive, clearly z is outright minimal among sets having the property $\mathcal{P}$.

Now we show that (3) is equivalent to (1), *assuming AC and Infinity.*

Theorem 1.7. *Assume Infinity.*

(a) *If (1) then (3).*
(b)^{AC} *If (3) then (1).*

Both proofs are easy. However, the proof of (b) involves an application of AC and a definition by induction. (See Problem 4.)

Problems

1. Prove 1.1 (a), (b), (c).
2. Prove 1.2.
3. Give a detailed proof of 1.5 including a proper definition by induction.
4. Prove 1.7(a) and (b)^{AC}.

§ 2. Consequences of regularity

We now adopt Reg; so we are working in $Z_0F + Reg$.

Using *Reg* we can now answer a question everyone asks as he begins to study set theory: Can a set belong to itself?

Theorem 2.1. *For any x,* $x \notin x$. *Also the following are impossible:* $x \in y \in x$; $x \in y \in z \in x$; *etc., etc..*

Proof. 2.1 is clear from (1) (which follows from *Reg*). For example, if $x \in y \in z \in x$, then the set $\{x, y, z\}$ would have no minimal element.

In $Z_0F + Reg$ (but not in $ZF - Reg$, as is known) we have

Theorem 2.2. *AC implies* AC^+.

(AC^+ was stated in § 2 of Chapter 4.)

Proof. Suppose for any $i \in I$ there exists j such that $\mathcal{P}ij$. Let F be the function on I such that, for each $i \in I$, $F(i)$ is the set of all j such that $\mathcal{P}ij$ and j belongs to the $\subseteq$-least partial universe containing some j such that $\mathcal{P}ij$ (in other words j has the least possible rank among j' such that $\mathcal{P}ij'$). By its definition this set obviously exists. Clearly each $F(i) \neq \emptyset$. The choice function f that AC guarantees for F clearly is as desired in AC^+.

There are a number of uses of the Axiom of Regularity somewhat similar to that just given. A beautiful and very useful application of *Reg* of this type was discovered by Dana Scott [L1955]. It is given in 2.3 below.

2.3 provides a general solution of the 'abstraction problem'. This is the problem of finding for each predicate $\approx$, which is an 'equivalence' over a class $\mathcal{P}$, a corresponding notion of *type* such that for any x, y, if $\mathcal{P}x$ and $\mathcal{P}y$ then *type* x

$=$ *type* y if and only if $x \approx y$. (The abstraction problem was solved in Chapter 7, § 1 for isomorphism among well-orderings – by the method of von Neumann – and then for $\sim$ using AC (Chapter 7, § 4). In the same section it was also solved (more or less for general isomorphism) by the method of A. P. Morse. Scott's method is superior in that it works for all equivalences. Moreover, Scott's method does not depend (as Morse's does) on AC, or even Replacement (see § 3 below). But Morse's method, it must be said, does not depend on *Reg* (as Scott's does).)

Theorem 2.3 (*Scott*). *Suppose* $\approx$ *is an equivalence over* $\mathcal{P}$. *For any* x *such that* $\mathcal{P}x$, *put type* x = *the set of all* y *such that* $x \approx y$ *and* y *belongs to the least partial universe containing some* z *such that* $x \approx z$.

Suppose $\mathcal{P}x$ *and* $\mathcal{P}y$. *Then:*

$$\text{type } x = \text{type } y \text{ if and only if } x \approx y.$$

The reader can verify 2.3 in his head!

At the end of § 2 of Chapter 7, we had obtained Chapter 1 and § 1 and § 2 of Chapter 7, and had introduced von Neumann's ordinals – *all in* Z_0F. We then assumed AC and in § 4 got adequate notions of $\bar{\bar{A}}$ and $\mathrm{Tp}\underline{A}$. At this point, Chapters 2-5 were 'reabsorbed' (for the new $\bar{\bar{A}}$ and $\mathrm{Tp}\underline{A}$). But this was at the expense of obliterating all work in those chapters which established results without using AC!

Now we assume $Z_0F + Reg$ (rather than '$+ AC$' in § 4 of Chapter 7). Thus we have Chapter 1, § 1 and § 2 of Chapter 7, and Chapter 8 to date. We now apply Scott's 2.3 using $\sim$ (and later $\cong$) for $\approx$, and in this way define the new notions $\bar{\bar{A}}$ and $\mathrm{Tp}'\underline{A}$.

We next make some minor remarks and adjustments. Of course, we have $\bar{\bar{A}} = \bar{\bar{B}}$ if and only if $A \sim B$. By 2.2, AC implies AC^+ and hence (cf. § 2 of Chapter 4) AC implies Assumption 2.1 of Chapter 4. Just as in 7.4.6 we see that no $\mathrm{Tp}'A$ is ever a von Neumann ordinal. So we can define $\mathrm{Tp}\underline{A} = \mathrm{Ord}\,\underline{A}$ if $\underline{A}$ is well ordered and $\mathrm{Tp}\underline{A} = \mathrm{Tp}'\underline{A}$ otherwise. Obviously the assumptions of Chapter 5 are now met.

Thus *we have all of* Chapters 2-5 *for our new* '$\bar{\bar{A}}$' *and* '$\mathrm{Tp}\underline{A}$' (and now without destroying all distinctions concerning the use of AC).

In particular we have from Chapter 2 the notion 'finite' and many facts about it. We now agree to call x ***hereditarily finite*** (or, say, ***hereditarily green***) if for every $y \in \mathrm{Tc}(\{x\})$, y is finite (or green).

Theorem 2.4. x *is hereditarily finite if and only if for some* n, $x \in V_n$.

Proof. See Problem 1. This is straightforward but takes some effort. (In one direction, use 'induction on $\mathrm{Rk}\,x$'.)

Remark. We could have discussed 'finite' originally in Chapter 2 without using cardinal numbers (thus avoiding 2.3 in proving 2.4 above.) But this would have been a considerable bother.

We say that a transitive family W of sets has *class replacement if* $F[A] \in W$ whenever $A \in W$ and $F\colon W \to W$.

Theorem 2.5. *V_ω has class replacement, if it exists.*

Proof. Let $F\colon V_\omega \to V_\omega$ and $A \in V_\omega$. Thus for some n, $A \in V_n$, so by 2.4, A is finite; and hence $F[A]$ is finite. Thus, $\{\mathrm{Rk}\, x\colon x \in F[A]\}$ is finite, so has a largest member k. Thus $F[A] \subseteq V_{k+1}$ and hence $F[A] \in V_\omega$, as desired.

Using AC, we generalize 2.5 and also establish a converse:

Theorem 2.6^{AC}. *An ordinal γ is an inaccessible cardinal if and only if V_γ has class replacement.*

Proof. (→) Assume γ is an inaccessible cardinal, say, κ. κ is a limit ordinal (like any infinite cardinal). We show by induction (on 'α') that for all $\alpha < \kappa$,

(1) $\overline{\overline{V_\alpha}} < \kappa$. (See Problem 2.)

Using (1) it is not difficult to show

(2) *V_κ has class replacement.* (See Problem 3.)

Now assume V_γ has class replacement. Note that $\alpha \in V_\gamma$ if and only if $\alpha < \gamma$. We claim that

(3) If $x \in V_\gamma$ then $\overline{\overline{x}} < \overline{\overline{\gamma}}$.

Indeed, if $\overline{\overline{x}} \geq \overline{\overline{\gamma}}$, then there exists $y \subseteq x$ such that $y \sim \gamma$. Clearly $y \in V_\gamma$ so by class replacement, $\gamma \in V_\gamma$ – which is absurd.

Using (3), we see at once that γ is a cardinal, say, κ. If $\lambda < \kappa$ then (since κ is a limit ordinal) $P(\lambda) \in V_\kappa$; so, by (3), $2^\lambda < \kappa$. Finally, κ is regular. If not, let A be a cofinal subset of κ of power $\lambda < \kappa$. By class replacement, $A \in V_\kappa$. Thus $\mathrm{Rk}\, A < \kappa$; but since A is cofinal in κ, $\mathrm{Rk}\, A = \kappa$, a contradiction.

Problems

1. Prove 2.4.
2. Prove (1).
3. Prove (2).

§ 3. Avoiding Replacment

As Scott observed, his definition (in 2.3) works even for *ordinals* in an important situation where von Neumann's fails – to wit, if the Axiom of Replacement is dropped.

Replacement is perhaps the most dubious axiom of *ZFC*, since it seems at least possible that the rest of the axioms are consistent, but become inconsistent when Replacement is added. Moreover, Replacement is almost never needed in ordinary mathematics (which rarely goes above $V_{\omega+\omega}$).

Actually, it is a routine affair to go back through all of this book to date and determine which results can easily be obtained without Replacement. (Many can, but some cannot.) We could have done this all along, but it would have obscured more important matters. This section will be a little sketchy, as we will leave it to the reader to verify that certain simple things can be done without Replacement.

We now work in Z_0 + Infinity = $Z - Reg$. Thus § 1, as it stands, is not available, since the opening recursive definition of the V_α requires Replacement. Indeed, even the axiom *Reg* makes no sense. The alternatives (1) and (2) of § 1 do make sense, and are often 'mistakenly' taken here as the axiom of regularity, but they are too weak. The 'right' version of *Reg* will appear below.

By a *tower* we mean any family t of sets such that

(i) t is a chain and every non-empty subset of t has an $\subseteq$-least member (in other words, t is well-ordered by $\subseteq$),

(ii) whenever W is the immediate predecessor of U (in our chain), then $U = P(W)$, and

(iii) whenever $U \in t$ has no immediate predecessor, then

$$U = \bigcup_{\substack{W \in t \\ W \subsetneq U}} W.$$

The key fact here is

Theorem 3.1. *If t and t' are towers, then one* is *an initial segment of the other.*

Proof. This is an easy consequence of 7.1.2 (Comparability of well-orders) (see Problem 1). (The old proof of 7.1.2 works without Replacement.)

By 3.1, all towers are 'stacked'.

Call W a *partial universe* if for some tower t, $W \in t$. More or less as in § 1, one easily shows, *using* 3.1:

Proposition 3.2.

(a) *Any partial universe is supertransitive.*

(b) *The class of partial universes is strictly ordered by $\subseteq$ or (what is here the same) $\in$.*

(c) *The predecessors of any partial universe W form a set.*
(d) *If for some partial universe W, $\mathcal{P}W$, then there is an $\subseteq$-least such W.*

Of course, a partial universe is called a limit partial universe if it is a limit element in the 'order' of partial universes. It is very easy to check

Proposition 3.3. *Assume Replacement. Then the new notions 'partial universe', and 'limit partial universe' agree with the old (from § 1).*

In Z_0 + Infinity we 'again' take for *Reg* the statement: *Every set belongs to some partial universe.*

Now, adopting *Reg*, we see at once that

Theorem 3.4. *Scott's Theorem 2.3 holds verbatim.*

Here is an example: Consider a world in which the sets in $V_{\omega+\omega}$ are the only sets. Clearly all axioms of *Z* hold (including *Reg* as above); but Replacement fails. Von Neumann's definition of ordinal is not available, even though *AC* holds, but Scott's is!

Problem

1. Prove 3.1.

9. LOGIC AND FORMALIZED THEORIES

Our discussion in this chapter will not be within set theory, about sets. Rather it will be about expressions, proofs, etc. Our discussion will therefore lie within what is called *metamathematics*.

In § 1-3 we describe a typical formalized theory. We do not assume yet that it is a 'set theory'. Otherwise, what we do would appear to be special for set theories, when it is really quite simple and general. So our theory might be a theory of ordered fields, or a number theory, or a set theory, or whatever.

Before really beginning this chapter in § 1 below, we make some informal remarks about our 'old' language – the one described in § 1 of Chapter 6. When we *use* a language (as in Chapters 1-8) it is convenient to have a rather large, redundant list of logical symbols. On the other hand, when (as below) we theorize *about* a language, the opposite is true. So it will be useful to note (informally) that our old language would have been just as rich (i.e., could have *said* the same things) if it had had only $\sim$, $\rightarrow$, $\forall$, and $=$ (and not $\wedge$, $\vee$, $\leftrightarrow$, $\exists$, $\exists!$, or ι), and also had had no operator variables. Indeed we can obviously replace $\mathcal{P} \vee \mathcal{Q}$ by $\sim \mathcal{P} \rightarrow \mathcal{Q}$; $\mathcal{P} \wedge \mathcal{Q}$ by $\sim(\sim \mathcal{P} \vee \sim \mathcal{Q})$, $\mathcal{P} \leftrightarrow \mathcal{Q}$ by $(\mathcal{P} \rightarrow \mathcal{Q}) \wedge (\mathcal{Q} \rightarrow \mathcal{P})$, $\exists x \mathcal{P}x$ by $\sim \forall x \sim \mathcal{P}x$ and $\exists! x \mathcal{P}x$ by $\exists x \forall y (\mathcal{P}y \leftrightarrow y = x)$. Next, we can omit, say, a two-place operator symbol $\mathcal{A}$ by taking an unused 3-place $\mathcal{P}$ and always using in place of $\mathcal{A}_{xy}$ the expression $\iota z \mathcal{P}xyz$. Finally, using convention 6.1.1 (for handling improper descriptions) one can see in his head that, in an awkward but obvious way: if we are given any asserting expression Σ involving ι we can construct an equivalent one not involving ι. For example, suppose Σ is

$$\iota x \Gamma x \in A.$$

We first form

$$\exists y (y = \iota x \Gamma x \wedge y \in A).$$

Then by 6.1.1 we can replace the part $y = \iota x \Gamma x$ by

$$(\exists! x \Gamma x \wedge \Gamma y) \vee (\sim \exists! x \Gamma x \wedge \forall u(u \notin y)).$$

In any case, below we shall only consider languages having the smaller list of symbols. If we speak below of an axiom, say, of *ZF*, we mean the axiom has been modified as above. Just for the record, here is how some of the modified axioms might look, still using $\wedge$, $\exists$, etc., but not ι:

Separation Axiom. $\exists B \forall x(x \in B \leftrightarrow x \in A \wedge \mathcal{P}x)$.

Doubleton Axiom. $\exists A \forall u(u \in A \leftrightarrow u \approx x \vee u \approx y)$.

Union Axiom. $\exists Z \forall x(x \in Z \leftrightarrow \exists A(x \in A \wedge A \in Z))$.

Power Set Axiom. $\exists U \forall A(A \in U \leftrightarrow \forall x(x \in A \rightarrow x \in B))$.

Replacement Axiom. $\forall x \exists! y \mathcal{P}xy \rightarrow \exists B \forall y(y \in B \leftrightarrow \exists x(x \in A \wedge \mathcal{P}xy))$.

§ 1. Language and grammar

First we describe precisely a typical formalized language $\mathfrak{L}$. The language $\mathfrak{L}$ has a certain *vocabulary* (special for $\mathfrak{L}$), which is just a collection of relation symbols each having a certain number of places. '$\mathcal{P}$' and '$\mathcal{Q}$' will now always denote relation symbols *of* $\mathfrak{L}$. $\mathfrak{L}$ has the following other symbols:

Left and right parentheses (,).
Special logical symbols $\sim$, $\rightarrow$, $\forall$, $\approx$.
Ordinary variables $x, y, \ldots, A, B, \ldots$.

An *expression* Γ (of $\mathfrak{L}$) is any non-empty concatenation of symbols, say, $\sigma_0 \sigma_1 \cdots \sigma_n$. The *length* $\ell(\Gamma)$ of Γ is $n + 1$. Γ' is a *segment* of Γ if it is of the form $\sigma_i \sigma_{i+1} \cdots \sigma_j$; an *initial segment* if also $i = 0$; a *proper segment* if also $\Gamma' \neq \Gamma$. Obviously we can define a notion Γ^* for all expressions Γ by a kind of recursion in which, when we define Γ^*, we take Δ^* as already known for every proper segment Δ of Γ. There is a corresponding kind of induction.

We now give (for the first time) a *precise* definition of 'asserting expressions' and 'naming expressions'. As is common in metamathematics, they will be referred to instead as *formulas* and *terms* respectively. Because we do not have now the ι-symbol, the terms are just the ordinary variables. An expression Γ is an atomic formula if it is of the form $\mathcal{P}x$ ($\mathcal{P}$ one place) or $\mathcal{Q}xy$ ($\mathcal{Q}$ two places), etc., or else Γ is of the form $\approx xy$. We now give a recursive definition (as above) of being a formula. Γ is a formula if it is an atomic formula or else is of the form $\sim \Delta$, $(\Delta \rightarrow \Delta')$, or $\forall x \Delta$, where Δ and Δ' are formulas. The letters 'Θ', 'Φ', and 'Ψ' will always denote formulas.

We want a formula to be readable in only one way. For our language this reduces at once to seeing that

(1) If $(\Phi \to \Psi) = (\Phi' \to \Psi')$ then $\Phi = \Phi'$ and $\Psi = \Psi'$,

(1) follows easily (see Problem 1) from

(2) If expression Γ is a formula then none of its proper initial segments is a formula.

(2) is proved (see Problem 1) by 'induction' on Γ (see above).

Aside. It is easy to see that (in theory) a machine could be built such that if any expression is put into it, it will after a while issue a piece of paper saying correctly 'yes, a formula' or 'no, not a formula'. Eventually we want the same to apply to the notion "is a proof".

The language $\mathcal{L}$ will never be modified. But in the metalanguage we will be rather casual in writing a *name* of a formula. Thus, we shall be normally careless about parentheses; we write $x \approx y$ for $\approx xy$, and we write $\exists x\Phi$ for $\sim \forall x \sim \Phi$, $\Phi \vee \Psi$ for $\sim \Phi \to \Psi$, etc., etc.

Another group of grammatical notions also mentioned briefly in Chapter 1 will now be described precisely – namely, 'free', 'bound', etc. We define by recursion on Φ the notion: in Φ, a certain occurrence of x is bound by a certain occurrence of $\forall$ – as follows:

If Φ is atomic, no occurrence of x is bound by an occurrence of $\forall$.

If Φ is $(\Psi_1 \to \Psi_2)$, an occurrence of x is bound by an occurrence of $\forall$ just if both occurrences lie in the same one of Ψ_1 and Ψ_2 and there the (corresponding) occurrence of x is bound by the (corresponding) occurrence of $\forall$.

Similarly if Φ is $\sim \Psi$.

If Φ is $\forall y \Psi$, then in Φ an occurrence of x is bound by one of $\forall$ if and only if both are in Ψ and there the corresponding occurrence of x is bound by that of $\forall$; or y and x are the same, the occurrence of $\forall$ is at the first place in Φ and either the occurrence of x is at the second place in Φ or else it lies in Ψ and is not bound in Ψ by any occurrence in Ψ of $\forall$.

Several other notions can be defined directly now. An occurrence of x in Φ is called *bound* if it is bound by some occurrence of $\forall$; it is called *free* if it is not bound. x (not an occurrence of x) is called a *free* (or *bound*) *variable of* Φ if it occurs free (or bound, respectively) in Φ at least once. A formula with no free variables is called a *sentence*. 'Σ' will always denote a sentence.

Φ is called a *universalization* of Ψ if Φ is of the form $\forall x_1 \cdots \forall x_n \Psi$.

The language we have described is called the language of *first-order logic* or *predicate logic*.

Problems

1. Prove (2).
2. Derive (1) from (2).

§ 2. Truth Tables

The part of logic which ignores $\forall$ and looks only at $\sim$ and $\rightarrow$ (called sentential logic) is very simple. Suppose we partly decompose a formula Φ, always stopping if we get to an atomic formula or one of the form $\forall x \Psi$, and possibly stopping anyway. Replace the subformulas stopped at by new symbols $\mathfrak{p}$, $\mathfrak{q}$, $\mathfrak{r}$, etc. (identical subformulas being replaced by the same symbol.) This gives us a special 'sentential' formula Γ, related to Φ. For example Γ might be

(1) $$\mathfrak{p} \rightarrow (\sim \mathfrak{p} \rightarrow \mathfrak{q}).$$

Writing T for 'true' and F for 'false' the *truth tables* for $\sim \mathfrak{p}$ and $\mathfrak{p} \rightarrow \mathfrak{q}$ are:

$\mathfrak{p}$	$\sim \mathfrak{p}$
T	F
F	T

and

$\mathfrak{p}$	$\mathfrak{q}$	$\mathfrak{p} \rightarrow \mathfrak{q}$
T	T	T
T	F	F
F	T	T
F	F	T

In the obvious way we construct the *truth tables* for any compound sentential formula. For example, (1), above, has the table

$\mathfrak{p}$	$\mathfrak{q}$	$\sim \mathfrak{p}$	$\sim \mathfrak{p} \rightarrow \mathfrak{q}$	$\mathfrak{p} \rightarrow (\sim \mathfrak{p} \rightarrow \mathfrak{q})$
T	T	F	T	T
T	F	F	T	T
F	T	T	T	T
F	F	T	F	T

An arbitrary formula Φ of $\mathfrak{L}$ is called a (*truth table*) *tautology* if there is a partial decomposition of Φ, and a related Γ, as above, such that Γ has a truth table whose last column is all T's. For example,

$$\forall x(x < z) \rightarrow (\sim \forall x(x < z) \rightarrow (y = z \vee y < z))$$

is a tautology, since (1) above is a Γ for it, and the last column of the table above is indeed all T's. A formula Ψ is called a *tautological consequence of* $\Phi_1, \ldots, \Phi_N$ if $\Phi_1 \wedge \ldots \wedge \Phi_N \rightarrow \Psi$ is a tautology.

Roughly speaking, a tautological formula is just one which is always valid simply because of the way $\sim$ and $\rightarrow$ appear in it. The truth table provides a machine-like method for determining whether a formula is such. In the full (predicate) logic one wishes to discuss formulas which are always valid *because of the way* $\sim$, $\rightarrow$, $=$, $\forall$ *and variables appear in them.* It is known that no machine-like method exists for picking out these formulas! This is a beautiful result of church [1936] which depends on the famous result of Gödel [1931]. The best we can hope to do is to introduce an adequate and machine-like notion of 'proof' (as in the next section). We will then know whether a proposed proof of a sentence is a proof or not, but, if it is not, there may be others, and we cannot examine all possible ones.

§ 3. Formal proofs

A formula (of our fixed $\mathcal{L}$) is called a *logical axiom* if it is a universalization of a formula of one of the following forms:

(I) A tautology,
(II) $\forall x(\Phi \to \Psi) \to (\forall x\Phi \to \forall x\Psi)$,
(III) $\Phi \to \forall x\Phi$ (where x is not free in Φ),
(IV) (a) $\exists x\, x = y$ (where x and y are distinct)
(b) $x = x' \to (\Phi \leftrightarrow \Phi')$ (where Φ is an *atomic* formula and Φ' results from Φ by replacing one occurrence of x by x').

Φ is said to be obtained from Θ_1 and Θ_2 by (the rule of) *Modus Ponens* if one Θ_i is of the form $\Psi \to \Phi$ while the other Θ_i is Ψ.

Let Ax be (always) a set of *sentences*. (We may think of Ax as a set of non-logical axioms for a theory. If Ax is empty, we are back in pure logic.) By a *proof of a formula Φ from Ax* we mean a finite sequence $\Psi_0, \cdots, \Psi_n$ of formulas such that $\Psi_n = \Phi$ and, for each $i \leq n$, either Ψ_i belongs to Ax (i.e., is a non-logical axiom) or Ψ_i is a logical axiom, or else Ψ_i is obtained from earlier Ψ_j's by Modus Ponens.

Φ is said to be a *consequence of Ax* (or a *theorem of Ax*), and we write $Ax \vdash \Phi$, if there is proof of Φ from Ax. (Of course, $\vdash \Phi$ means $\emptyset \vdash \Phi$). Other related terminology: Ax is *consistent* if for some Φ, $Ax \nvdash \Phi$ (or, what is the same, if, say, $Ax \nvdash x \neq x$), Σ is *independent of Ax* if $Ax \nvdash \Sigma$ (or, what is the same, if $Ax \cup \{\sim \Sigma\}$ is consistent. (Note that, 'independent' and 'consistent' are almost redundant – one could get along perfectly well with just one of the two!) Φ and Ψ are *Ax-equivalent* if $Ax \vdash \Phi \leftrightarrow \Psi$.

Aside. Most of the ideas discussed so far in this section appeared first in Frege [1879].

We may try to compare our formal system with the totality of logical procedures (axioms and rules) which are generally accepted in proofs in ordinary mathematical discourse. A short glance above will show that our procedures are among those commonly accepted. Many other procedures are also commonly accepted. Can each of these be duplicated in our system? Several such results appear below, e.g., in 3.4–3.7, and many more can be found in books on logic. From all these results it appears very, very likely that every formula whose logical validity can be established by an acceptable logical argument from ordinary mathematics can in fact be proved in the formal system above. This observation or informal conjecture probably goes back to Frege in some form. Today, a mathematical author, wishing to show that $Ax \vdash \Phi$ (for a certain Ax and Φ), commonly just gives the corresponding proof in informal mathematics – counting on the 'fact' that such a proof can be routinely translated into a formal proof.

A different and stronger conjecture, which (unlike that above) is precise, was made by Hilbert and Ackermann [1928], and established by Gödel [1930] in his dissertation. This important result is known as Gödel's completeness theorem. Roughly speaking, it says that if a sentence Σ is 'true' in all possible 'models' then Σ has a formal proof (i.e., $\vdash \Sigma$). However, in order to prove this result, or even to make its statement precise by defining 'true' and 'model' (*infinite* models must be allowed), we must use set theory in our metatheory. All of the other metadiscussions anywhere in this book can be based on much weaker, 'finitary' reasoning. We shall not make use of the completeness theorem.

We now begin to develop our system of logic. If Φ is a tautological consequence of $\Psi_1, \cdots, \Psi_n$ and $\Sigma \vdash \Psi_i$ (for $1 \leq i \leq n$) then (by Modus Ponens and Axiom I) clearly $\{\Sigma\} \vdash \Phi$. In such a case we will just say: "thus $\{\Sigma\} \vdash \Psi$ by *tautological reasoning*.

Metatheorem 3.1. (*The derived rule of inference called 'Generalization'.*) *If* $\vdash \Phi$ *then* $\vdash \forall x\Phi$.

Proof. Let $\Phi_0, \cdots, \Phi_n$ be a proof of Φ. It will be enough to show by induction on 'i' that for each $i \leq n$, $\vdash \forall x\Phi_i$. Assume this is so for all $i' < i$. If Φ_i is an axiom obviously so is $\forall x\Phi_i$. Otherwise Φ_i is obtained by Modus Ponens. Say Ψ and $\Psi \rightarrow \Phi_i$ are both earlier Φ_i's. Then by the inductive hypothesis, $\vdash \forall x\Psi$ and $\vdash \forall x(\Psi \rightarrow \Phi_i)$. Using Axiom II and tautological reasoning, we get $\vdash \forall x\Phi_i$, as desired.

It may be worthwhile to state in general a method justified by the form of the preceding argument. To show that every theorem from Ax has a certain property (*), it is enough to show that all members of Ax have property (*); all logical axioms have (*); and whenever Φ and $\Phi \rightarrow \Psi$ both have (*) so has Ψ.

The next theorem reduces the study of derivability from non-logical axioms to that of derivability in pure logic.

Metatheorem 3.2. (The Deduction Theorem.) (a) $\{\Sigma\} \vdash \Phi$ *if and only if* $\vdash \Sigma \rightarrow \Phi$ (b) $Ax \vdash \Phi$ *if and only if for some finite conjunction* Σ *of members of* Ax, $\vdash \Sigma \rightarrow \Phi$.

Proof. (b) follows easily from (a); and ($\leftarrow$) in (a) is immediate, by Modus Ponens. Consider ($\rightarrow$) in (a). We want to show that every theorem Φ from $\{\Sigma\}$ has the property that $\vdash \Sigma \rightarrow \Phi$. But, in the first place, if Φ is Σ, then we have $\vdash \Sigma \rightarrow \Sigma$, as $\Sigma \rightarrow \Sigma$ is a tautology. Secondly, suppose Φ is a logical axiom. Then $\vdash \Phi$ so, by tautological reasoning, $\vdash \Sigma \rightarrow \Phi$, as desired. Finally, suppose $\vdash \Sigma \rightarrow \Phi_1$

and $\vdash \Sigma \rightarrow (\Phi_1 \rightarrow \Phi_2)$. Then, by tautological reasoning, $\vdash \Sigma \rightarrow \Phi_2$. Thus $\vdash \Sigma \rightarrow \Phi$ for all theorems Φ from $\{\Sigma\}$, completing the proof of 3.2.

The rule 3.1 now extends easily to any set Ax (but only because we always insist that all members of Ax be *sentences*.)

3.3. *If $Ax \vdash \Phi$ then $Ax \vdash \forall x\Phi$.*

Proof. It is enough to take $Ax = \{\Sigma\}$. Let $\{\Sigma\} \vdash \Phi$. By 3.2, $\vdash \Sigma \rightarrow \Phi$. By 3.1, $\vdash \forall x(\Sigma \rightarrow \Phi)$. By Axiom II and tautological reasoning, $\vdash \forall x\Sigma \rightarrow \forall x\Phi$. By Axiom III, $\vdash \Sigma \rightarrow \forall x\Sigma$ (since Σ is a sentence). Thus we get $\vdash \Sigma \rightarrow \forall x\Phi$ and $\Sigma \vdash \forall x\Phi$ as desired.

Now we show that certain special (types of) sentences are logical theorems. The first result generalizes Axiom IV (b) in two ways.

Proposition 3.4. $\vdash x = y \rightarrow (\Phi \leftrightarrow \Phi')$, *if Φ' can be obtained from Φ by replacing one or more free occurrences of x by* free *occurrences of y.*

Proof. We can take x and y to be different (since otherwise our formula is a tautology). It is easily seen to be enough to deal with the replacement of *one* occurrence of x, since the general case can be obtained by several of these in succession. We prove the simplified 3.4 by induction on the formation of the formula Φ. If Φ is atomic, the desired result is a case of Axiom IV (b). The case of $\sim$ or $\rightarrow$ is easy using only tautological reasoning. Suppose Φ is $\forall z\Psi$ and the hypothesis of 3.4 holds for Φ and Φ'. Φ' can be written $\forall z\Psi'$. By the hypothesis about Φ and Φ', z must not be x or y. So clearly, Ψ' is obtained from Ψ by replacing one free occurrence of x by a free occurrence of y. Hence by the inductive hypothesis, $\vdash x = y \rightarrow (\Psi \leftrightarrow \Psi')$. By (3.1), Axiom II, and tautological reasoning, $\vdash \forall z(x = y) \rightarrow (\forall z\Psi \leftrightarrow \forall z\Psi'))$. But by Axiom III, $\vdash x = y \rightarrow \forall z\,(x = y))$. So $\vdash x = y \rightarrow(\Phi \leftrightarrow \Phi')$, as desired.

We write $\Phi(^x_y)$ for the result of substituting y for each *free* occurrence of x in Φ. By induction on Φ, $\Phi(^x_y)$ is easily seen to be a formula. More generally, $\Phi(^{x_1 \ldots x_k}_{y_1 \ldots y_k})$ (where the x_i are distinct) will denote the result of simultaneously substituting y_1 for each free occurrence of x_1 in $\Phi, \ldots,$ and y_k for each free occurrence of x_k in Φ.

Proposition 3.5. $\vdash \forall x\Phi \rightarrow \Phi(^x_y)$ *provided that y is free in $\Phi(^x_y)$ at each place where x was free in Φ.*

Note. The need for some restriction in 3.5 is clear from the following example: Let Φ be $\exists y\,(x < y)$. Then $\forall x\Phi \rightarrow \Phi(^x_y)$ is $\forall x\exists y\,(x < y) \rightarrow \exists y\,(y < y)$ – clearly not logically valid! One says in such a case that there is a 'collision of variables'.

Proof. First we will show that (*) 3.5 *holds if x and y are different variables.* By 3.4 we have $\vdash x = y \to (\Phi \leftrightarrow \Phi(\tfrac{x}{y}))$, as the hypothesis of 3.4 is implied by that of 3.5. So easily we get

$\vdash \forall x\Phi \to \forall x\,(x = y \to \Phi(\tfrac{x}{y}))$; and then (using contrapositives)
$\vdash \forall x\Phi \to \forall x\,({\sim}\Phi(\tfrac{x}{y}) \to x \neq y)$ and
$\vdash \forall x\Phi \to (\exists x\,(x = y) \to {\sim}\forall x{\sim}\Phi(\tfrac{x}{y}))$.

Since x and y are different, x is not free in ${\sim}\Phi(\tfrac{x}{y})$. So by Axiom III, $\vdash {\sim}\Phi(\tfrac{x}{y}) \to \forall x{\sim}\Phi(\tfrac{x}{y})$. Thus $\vdash \forall x\Phi \to (\exists x\,(x = y) \to {\sim}{\sim}\Phi(\tfrac{x}{y}))$. By Axiom IV(a) (since x and y are different) and tautological reasoning, we get $\vdash \forall x\Phi \to \Phi(\tfrac{x}{y})$. Thus (*) is proved.

Finally, we must still show that $\vdash \forall x\Phi \to \Phi$. Let z be 'new'. By (*), $\vdash \forall x\Phi \to \Phi(\tfrac{x}{z})$. Using 3.1, and Axioms III and II we easily get $\vdash \forall x\Phi \to \forall z\Phi(\tfrac{x}{z})$. By (*) again, $\vdash \forall z\Phi(\tfrac{x}{z}) \to \Phi(\tfrac{x}{z})(\tfrac{z}{x})$ (as the hypothesis of 3.5 clearly holds here). But $\Phi(\tfrac{x}{z})(\tfrac{z}{x})$ is just Φ, so $\vdash \forall x\Phi \to \Phi$, as desired.

Incidentally, of the three 'missing' familiar axioms $x = x$, $x = y \to y = x$, and $x = y \wedge y = z \to y = z$, the second and third are really cases of IV(b). That $\vdash x = x$ is a small exercise. (Problem 1.)

We have not taken 3.5 as an axiom, as is very commonly done. Our particular system comes from one due to Tarski [1951], as improved by Kalish and Montague [1956]. Its special feature is just that its axioms do not involve the notion of general substitution $\Phi(\tfrac{x}{y})$. This will be quite expedient for us below in § 4 and § 5. From 3.1-5 we have the validity in our system of all the axioms and rules found in nearly any book.

Intuitively, it is clear that bound variables can be changed in a formula, giving an equivalent formula. Precisely, we call Φ and Φ' *(alphabetic) variants* if they have the same length, and, at each place, either they have the same symbol not a variable, or they both have a free occurrence of the same variable, or else they both have a bound occurrence of a variable (not necessarily the same one) which is bound by an occurrence of $\forall$ at the same place in each.

Proposition 3.6. $\vdash \Phi \leftrightarrow \Phi'$, *for any variant Φ' of Φ.*

The proof of 3.6 is in the Appendix. In the remainder of Chapter 9 we shall omit the proofs of 3.6, 4.1, and 5.1 – which are the only proofs more than a few lines long. These omitted proofs are in the Appendix and can be read by readers who wish to, *as they go along in the rest of Chapter 9.* A number of the proofs in logic or metamathematics (such as those of 4.1 and 5.1 below) involve a lot of detail, although they are not really difficult. We suggest to readers who feel impatient that they just learn what the facts are, by reading the rest of Chapter 9, which omits these longer proofs. They can then move on with no difficulty to Chapter 10 on independence proofs, which combines the methods of set theory (Chapters 1-8) and metamathematics (Chapter 9). Up to now in Chapter 9, all proofs have been given. Thus the reader will be able to grasp fully and exactly what the definitions and metatheorems in the rest of Chapter 9 *say*. Perhaps after reading Chapter 10 some more readers will wish to read the missing proofs given in the Appendix.

§ 4. Substitution for predicate symbols and relativization.

We now discuss a key matter, the substitution of a formula Θ for a predicate symbol $\mathcal{P}$ in a formula Φ. Suppose $\mathcal{P}$ is k-ary. Actually the thing to substitute for $\mathcal{P}$ is not just a formula Θ, but a tuple $(\Theta; x_1, \cdots, x_k)$ where the x_i are distinct variables. (Roughly speaking, such a tuple creates or denotes a k-ary predicate when *other* variables (than the x_i's) free in Θ are kept fixed or treated as parameters, while $x_1, \cdots, x_k$ are 'varied'). We write

$$\Phi\begin{bmatrix}\mathcal{P}\\(\Theta; x_1, \ldots, x_k)\end{bmatrix}$$

for the formula obtained from Φ by replacing each occurrence of an atomic formula starting with $\mathcal{P}$, say $\mathcal{P}y_1\cdots y_k$, by $\Theta\binom{x_1\cdots x_k}{y_1\cdots y_k}$. To simplify notation we also write $\Phi(S)$, where S is the 'substitution' (or 'substitution instruction'): $\binom{\mathcal{P}}{(\Theta; x_1, \cdots, x_k)}$.

There is a corresponding notion of simultaneous substitution. A 'substitution' is now a "monster":

$$(1) \qquad S = \begin{bmatrix}\mathcal{P}_1 & \cdots & \mathcal{P}_m\\(\Theta_1; x_{11}, \ldots, x_{1\ell_1}) & \cdots & (\Theta_m; x_{m1}, \ldots, x_{m\ell_m})\end{bmatrix}.$$

(Here the $\mathcal{P}_i$'s are distinct, $\mathcal{P}_i$ is ℓ_i-ary, and, for each i, $x_{i1}, x_{i2}, \ldots$ are distinct. Sometimes we treat S as a mapping defined on $\{\mathcal{P}_1, \ldots, \mathcal{P}_m\}$.) *We denote by* $\Phi(S)$ *the formula obtained from* Φ *by simultaneously replacing, for* $i = 1, 2, \ldots, m$, *each occurrence of an atomic formula, say* $\mathcal{P}y_1 \ldots y_{\ell_i}$, *by* $\Theta_i\binom{x_{i1}\cdots x_{i\ell_i}}{y_1\cdots y_{\ell_i}}$. Obviously, the operator — (S) goes through $\sim, \rightarrow$, and even $\forall w$. That is, for example, $(\forall w\phi)(S) = \forall w\Phi(S)$. Just to save time, let us agree that henceforth, *whenever the letter 'S' is used, we assume* that (1) *holds*.

Obviously in forming $\Phi(S)$ there is danger that collisions of variables will occur. Let us say that z is *a variable of S* (or *in S*) if for some i, z is a variable in Θ_i (free or bound), or else z is one of the x_{ij}'s. We say that S is *safe* for Φ (our 'safe' is really 'very safe') if S and Φ have no common variables.

Let us write $\vdash^{\mathcal{L}'}\Phi$ to mean that Φ has a proof (lying entirely) in the language $\mathcal{L}'$.

Metatheorem 4.1. (*Substitution for predicate variables*).
(a) *Assume that the substitution S is safe for* Φ, *and that* $\vdash\Phi$. *Then* $\vdash\Phi(S)$.
(b) *If* $\vdash\Phi$ *then* $\vdash^{\mathcal{L}'}\Phi$ *in any language* $\mathcal{L}'$ *containing* Φ *(as a formula)*.

The proof of 4.1 is in the appendix. As is seen there, the proof of (a) with small changes establishes (b) – which otherwise may not appear to be related to (a).

Regarding 4.1(b), consider a language $\mathcal{L}'$ having a smaller vocabulary than our fixed $\mathcal{L}$. (Φ, Θ_i, etc., above were all assumed to be in $\mathcal{L}$, and '$\vdash \Phi$' means $\vdash^{\mathcal{L}} \Phi$.) Let Γ be an expression of $\mathcal{L}'$. It is easy to see that Γ is a formula in the sense of $\mathcal{L}'$ if and only if it is a formula in the sense of $\mathcal{L}$ (and similarly for other grammatical notions like 'free variables'). However, it is not so clear that if the formula Γ of $\mathcal{L}'$ has a proof in $\mathcal{L}$ then Γ has a proof in $\mathcal{L}'$. Certainly the same proof need not work. From the intuitive notion of logical validity we would expect that Γ does indeed have a proof in $\mathcal{L}'$. 4.1(b) just establishes that this is so.

4.1 implies almost at once its own slight generalization 4.2: (A serious generalization is in 5.1.)

Corollary 4.2. *Let S be safe for Φ and suppose Ax is a set of sentences in which no $\mathcal{P}_i$ of S occurs. If $Ax \vdash \Phi$, then $Ax \vdash \Phi(S)$.*

Proof. By the Deduction theorem we may as well assume that $Ax = \{\Sigma\}$ and $\vdash \Sigma \rightarrow \Phi$. Since Σ has no free variables it is obvious we can find a variant Σ' of Σ such that S is safe for Σ', and hence for $\Sigma' \rightarrow \Phi$. By 3.6, $\vdash \Sigma \leftrightarrow \Pi'$, so $\vdash \Sigma' \rightarrow \Phi$. Applying 4.1, we have $\vdash \Sigma'(S) \rightarrow \Phi(S)$. But since no $\mathcal{P}_i$ of S is in Σ', $\Sigma'(S) = \Sigma'$; so clearly $\vdash \Sigma \rightarrow \Phi(S)$, as desired.

We turn now to the topic of relativization. Let $\mathcal{P}$ be a one-place relation symbol. By recursion on Φ we define below a formula $\Phi^{(\mathcal{P})}$ which, intuitively, says that Φ holds in the inner world of all things for which $\mathcal{P}$ holds. ($\Phi^{(\mathcal{P})}$ is read: Φ relativized to $\mathcal{P}$). If Φ is atomic, $\Phi^{(\mathcal{P})}$ is Φ. $(\sim \Phi)^{\mathcal{P}}$ is $\sim \Phi^{(\mathcal{P})}$ and $(\Phi \rightarrow \Psi)^{(\mathcal{P})}$ is $\Phi^{(\mathcal{P})} \rightarrow \Psi^{(\mathcal{P})}$. The only non-trivial condition is this: $(\forall x\Phi)^{(\mathcal{P})}$ is $\forall x(\mathcal{P}x \rightarrow \Phi^{(\mathcal{P})})$.

Metatheorem 4.3. *If* $\vdash \Sigma$ *then* $\vdash \exists x\mathcal{P}x \rightarrow \Sigma^{(\mathcal{P})}$. (*$\Sigma$ is any sentence.*)

Remarks. One can either allow empty 'worlds' or not. As is common, our axioms above are for a logic which only considers non-empty worlds. (See Problem 1.) In such a world there may, however, be no elements which satisfy $\mathcal{P}$. This explains the peculiar form of 4.3.

We could deal with $\Phi^{(\Psi)}$, where Ψ has one free variable, but it is more convenient to deal only with $\Phi^{(\mathcal{P})}$, as all collisions of variables are avoided.

Theorem 4.3 plays a key role in proofs of independence in set theory by the so-called method of inner models (worlds); cf., for example, Chapter 10.

Proof. (See Problem 2.) The proof is much simpler than that of 4.1. It is similar to the proofs of 3.1 and 3.2, though slightly more involved. In order to apply "induction along the proof", as there, we first need to replace 4.3 by an assertion about an arbitrary *formula* Φ (in place of Σ). We take this to be the statement: If $\vdash \Phi$, then

(*) if Φ has the distinct free variables $u_1, \ldots, u_k$, then $\vdash \exists x \mathcal{P}x \wedge \mathcal{P}u_1 \wedge \ldots \wedge \mathcal{P}u_k \rightarrow \Phi^{(\mathcal{P})}$.

The reader can complete the proof of 4.3 by showing that all logical axioms Φ satisfy (*) and that (*) is preserved by modus ponens.

Let Ψ be a formula whose only free variable is x. Combining 4.3 and 4.1, we can now deal adequately with relativization to Ψ (or the inner world corresponding to Ψ) – rather than $\mathcal{P}$. Let $\mathcal{P}$ be a fixed, one-place predicate symbol which is 'new' (not in our original language, $\mathcal{L}$). For any formula Φ of $\mathcal{L}$, let $\Phi^{*\Psi}$ (or just Φ^* when Ψ is fixed) be the formula $\forall x(\mathcal{P}x \leftrightarrow \Psi) \rightarrow \Phi^{(\mathcal{P})}$. If Ax and Ax' are sets of $\mathcal{L}$-sentences, we say Ax' *is interpreted in* Ax by Ψ if $Ax \vdash \exists x\Psi$ and for each sentence Σ in Ax', $Ax \vdash \Sigma^{*\Psi}$. (Of course, to say Ax' is interpret*able* in Ax means: – by some Ψ.) It is easy to see (*using* 4.3) that the collection of all (sentences) Σ such that $Ax \vdash \Sigma^{*\Psi}$ is 'deductively closed', that is:

Proposition 4.4. (a) *If* $Ax \vdash \Sigma_1^{*\Psi}$, ... *and* $Ax \vdash \Sigma_n^{*\Psi}$ *and if* $\vdash \Sigma_1 \wedge \ldots \wedge \Sigma_n \rightarrow \Sigma$, *then* $Ax \vdash \Sigma^{*\Psi}$. *(See Problem 3.)*

From (a) follows at once

4.4. (b) *If* Ax' *is interpreted in* Ax *by* Ψ *and* $Ax' \vdash \Sigma$, *then* $Ax \vdash \Sigma^{*\Psi}$.

We can now easily prove a very useful result, 4.5, about interpretation. 4.5 is the main thing needed from this chapter in Chapter 10.

Metatheorem 4.5. *If* Ax' *is interpretable in* Ax *and* Ax *is consistent, then* Ax' *is consistent.*

Proof. Assume Ax' is interpreted in Ax by Ψ. Obviously we can assume that Ax and Ax' are finite and even one-element sets. So we assume $Ax = \{\Sigma\}$ and $Ax' = \{\Sigma'\}$. Suppose Σ' is inconsistent. Let γ be a sentence involving only $\approx$ with all variables 'new'. Write f (for 'false') for the sentence $\gamma \wedge \sim \gamma$. We have $\Sigma' \vdash f$ and hence, by 4.4(b), $\Sigma \vdash \forall x(\mathcal{P}x \leftrightarrow \Psi) \rightarrow f^{(\mathcal{P})}$. But $f^{(\mathcal{P})}$ is $\gamma^{(\mathcal{P})} \wedge \sim \gamma^{(\mathcal{P})}$ so $\Sigma \vdash \sim \forall x(\mathcal{P}x \leftrightarrow \Psi)$. We would like to apply $\left(\begin{smallmatrix}\mathcal{P}\\(\Psi;\, x)\end{smallmatrix}\right)$, but it is not as 'safe' as required. So we pass to a variant Ψ' of Ψ whose bound variables are all 'new'. Also, we take a new variable x' and consider $S_0 = \left(\begin{smallmatrix}\mathcal{P}\\(\Psi'\binom{x}{x'},\, x')\end{smallmatrix}\right)$. Clearly S_0 is safe for $\sim \forall x(\mathcal{P}x \leftrightarrow \Psi)$. Hence, by 4.2, $\Sigma \vdash (\sim \forall x(\mathcal{P}x \leftrightarrow \Psi))(S_0)$. Now $(\mathcal{P}x)(S_0) = \Psi'\binom{x}{x'}\binom{x'}{x} = \Psi'$; so $\Sigma \vdash \sim \forall x(\Psi' \leftrightarrow \Psi)$. But, by 3.6, $\vdash \Psi' \leftrightarrow \Psi$, so Σ is inconsistent, as was to be proved.

Problems

1. Give without proof an example of a sentence (involving only =) which is a theorem of logic (as at the beginning of §3), but is not true in the empty world.

2. Complete the proof of 4.3.
3. Prove 4.4(a).

§ 5. Three forms of *ZFC*

To simplify the notation we now consider only 'set theories'. (But everything that we say can be extended at once to any other theories, like the usual formalized number theory, which have 'axiom schemas'.)

Let Ⓟ be a collection of relation symbols, containing infinitely many $\mathcal{P}$'s of each arity, and not containing the binary predicate symbol $\in$. The languages having respective vocabularies $\{\in\}$ and $\{\in\} \cup$ Ⓟ will be called $\mathcal{L}^{(0)}$ and $\mathcal{L}^{(1)}$. $\mathcal{L}^{(1)}$ will be taken as the primary language, so that, e.g., plain 'formula' means $\mathcal{L}^{(1)}$-formula.

By an *instance* of a formula Φ is meant any formula $\Phi(S)$, where S is a substitution which is safe for Φ and *S is not defined for* $\in$ (that is, no substitution is made for $\in$). The notion of instance was mentioned informally a few lines after 1.1.1. Note that:

(1) Any instance of Φ can be obtained as $\Phi(S)$, for another S, where in addition, the $\mathcal{P}_i$'s of S are exactly the relation symbols in Φ other than $\in$ and, for each i, $x_{i1}, x_{i2}, \cdots$ *do not occur bound in* Θ_i.

To see (1), drop $\mathcal{P}_i$'s not in Φ; send any $\mathcal{Q}$ in Φ missing in the old S to $(\mathcal{Q}u_1u_2\cdots; u_1, u_2\cdots)$; and change from $(\Theta_i; x_{i1}, \cdots)$ to $(\Theta_i(\begin{smallmatrix}x_{i1}\cdots\\ z_1\cdots\end{smallmatrix}); z_1, \cdots)$, where $z_1, z_2, \cdots$ are distinct and 'new'.

Let $\mathcal{Z}$ be a collection of sentences. $\mathcal{Z}$ induces two theories (i.e., axiom sets) $\mathcal{Z}^{(0)}$ and $\mathcal{Z}^{(1)}$. $\mathcal{Z}^{(1)}$ (in $\mathcal{L}^{(1)}$) consists of all sentences which are universalizations of instances of sentences in $\mathcal{Z}$. $\mathcal{Z}^{(0)}$ (in language $\mathcal{L}^{(0)}$) consists of those sentences in $\mathcal{Z}^{(1)}$ which are in $\mathcal{L}^{(0)}$. For example, $\mathcal{Z}$ might be *ZFC*. (We assume now that all axioms of *ZFC* have been universalized into sentences.) Ⓟ would then be the set of all predicate variables. Note that if $\mathcal{Z} = ZFC$ then $\mathcal{Z}$ would be finite, as Separation and Replacement are taken as single sentences involving predicate variables. (As in § 0, the '$\mathcal{A}$' in Replacement has become a binary $\mathcal{P}$.)

Using Metatheorem 4.1, we now establish in 5.1 some important facts about $\mathcal{Z}^{(1)}$ and $\mathcal{Z}^{(0)}$. The title of 5.1 sounds better in German (cf. Hilbert-Bernays [F 1934]).

Metatheorem 5.1. (On the putting back of substitutions.)
(a) *Suppose* $\mathcal{Z}^{(1)} \vdash \Phi$ *and* Ψ *is an instance of* Φ. *Then* $\mathcal{Z}^{(1)} \vdash \Psi$.
(b) $\mathcal{Z}^{(1)}$ *is a conservative extension of* $\mathcal{Z}^{(0)}$, *i.e., if* Φ *is in* $\mathcal{L}^{(0)}$ *and* $\mathcal{Z}^{(1)} \vdash \Phi$ *then* $\mathcal{Z}^{(0)} \vdash \Phi$.

The proof of 5.1 is in the Appendix.

Consider *ZFC* (but the same applies to *ZF*, Z_0F, etc.). By 5.1(b): *If*

$ZFC^{(1)} \vdash \Phi$ *where* Φ *involves only* $\in$, then $ZFC^{(0)} \vdash \Phi$. Hence, as long as we only consider formulas Φ, involving only $\in$ (which is so 98% of the time), there is no difference between $ZFC^{(0)} \vdash \Phi$ and $ZFC^{(1)} \vdash \Phi$. Thus the usual independence questions – like those considered in Chapter 10 or, e.g., "Is *CH* independent of *ZFC*?" are insensitive to which of $ZFC^{(0)}$ and $ZFC^{(1)}$ one has in mind.

Nevertheless $\mathcal{Z}^{(0)}$ and $\mathcal{Z}^{(1)}$ are obviously by no means identical. We now introduce a third formal system $\mathcal{Z}^*$ which appears to be quite new but is actually (essentially) identical to $\mathcal{Z}^{(1)}$. $\mathcal{Z}^*$ is not quite one of the systems we have studied up to now in this chapter, as it is not based on the same logic. By a **-proof of* Φ *from* $\mathcal{Z}$ we mean a finite sequence $\Phi_1, \cdots, \Phi_n$ of formulas (of $\mathcal{L}^{(1)}$) such that for each i, Φ_i is either a logical axiom (as in § 2) or is obtained from earlier Φ_j's by Modus Ponens, *or else* Φ_i *is a universalization of an instance of an earlier* Φ_j. (The last condition represents a new logical rule of inference.) We write $\mathcal{Z} \vdash^* \Phi$ to mean there is a *-proof of Φ from $\mathcal{Z}$. If we are working in the system $\mathcal{Z}^*$, we usually call the $\mathcal{P}$'s in Ⓟ predicate or relation *variables* (for obvious reasons). We consider the vocabulary of $\mathcal{Z}^*$ to be $\{\in\}$.

Corollary 5.2. $\mathcal{Z} \vdash^* \Phi$ *if and only if* $\mathcal{Z}^{(1)} \vdash \Phi$.

Proof. ($\leftarrow$) is trivial. ($\rightarrow$) follows at once from Theorem 5.1 because the only new thing in $\mathcal{Z}^*$, the new rule of inference, is a derived rule of inference is $\mathcal{Z}^{(1)}$ by 5.1 (and 3.1).

*The various *-systems –* Z_0F^*, ZFC^*, etc. – *are exactly what we have been working in (informally), in Chapters 1-8.*

Informal remarks. About which to use of two equivalent theories like $ZFC^{(1)}$ and ZFC^*, there can be no argument. One can even go back and forth between them if it is handy.

What about $ZFC^{(0)}$ versus ZFC^* (or $ZFC^{(1)}$)? $ZFC^{(0)}$ and ZFC^* have the same theorems involving only $\in$, so there is certainly not very much to choose between them. In recent years, it has been very popular in books and lectures to work in $ZFC^{(0)}$. We shall give one informal reason and one technical reason in favor of working in ZFC^*.

Informally, it seems obvious that a working mathematician (or a beginning student) would feel better about claiming he has proved in ZFC^* a single formula Φ like

"If there is an α such that $\mathcal{P}\alpha$ then there is a least α such that $\mathcal{P}\alpha$"

than he would about claiming that

(2) Every $\in$-instance of Φ is a theorem of $ZFC^{(0)}$.

The second argument depends on the following fact, due to Tarski: There is a formula Φ_0 (involving $\in$ and $\mathcal{P}$) such that (a) every $\in$-instance of Φ_0 is a theorem of $ZFC^{(0)}$, but (b) $ZFC \nvdash^* \Phi_0$. (Roughly speaking one can take for Φ_0 a formula saying "$\mathcal{P}$ is not a satisfaction predicate". Then (a) follows at once from the result of Tarski [1936] that 'satisfaction (or truth) is not definable within the system'.) Thus, to assert that $ZFC^{(1)} \vdash \Phi$ is in general *stronger* than claiming that every $\in$-instance of Φ is a theorem of $ZFC^{(0)}$. Now, the fact that (2) holds for Φ_0 has never appeared in a book on set theory! On the contrary, whenever (2) has appeared in such a book, the corresponding Φ has in fact been a theorem of ZFC^*. The argument in the book proving (2) in fact led to a proof that $ZFC \vdash^* \Phi$. In such a situation, if only (2) is claimed then some of the strength of the argument has been needlessly lost.

10. INDEPENDENCE PROOFS

In this chapter we shall prove four results, each having the form: If Ax is consistent, so is $Ax + \Sigma$ (that is, $\sim \Sigma$ is independent of Ax). In every case the method of proof will be, roughly speaking, to show that within any world in which Ax holds, one can create an inner world in which $Ax + \Sigma$ holds. This rough idea goes back at least to the models of non-Euclidean geometry (within models of ordinary Euclidean geometry) created by Beltrami (1868) and Klein (1871). In [1929], J. von Neumann established in such a way the relative consistency of the Axiom of Regularity; cf. Metatheorem 1.5 below.

§ 1. Relativization of axioms to partial universes. Consistency of adding *Reg*.

As usual, we write $(\forall x < y)\Psi$ for $\forall x(x < y \to \Psi)$. By recursion, we say that an $\in$-formula Φ is *bounded* if it is atomic or of one of the forms $\sim \Psi_1$, $\Psi_1 \to \Psi_2$, $(\forall x \subseteq y)\Psi_1$, or $(\forall x \in y)\Psi_1$, where Ψ_1 and Ψ_2 are bounded. (In other words, only $\in$-bounded or $\subseteq$-bounded quantifications are allowed in forming Φ).

Many familiar notions are bounded. For example one can easily show (see Problems 1 and 2) that each of the following formulas is equivalent in $Z_0^{(0)}$ to a bounded formula:

(1) $\begin{cases} \text{(a) '}u \subseteq v\text{' and also '}z = (x, y)\text{'; (b) '}w \text{ is a Peano structure'}; \\ \text{(c) '}y = P(x)\text{' (power set); and (d) '}z \text{ is a tower' (cf. Chapter 8, §2).} \end{cases}$

Naturally we write: '$\mathcal{P}$ is V' to mean $\forall x(\mathcal{P}x \leftrightarrow V(x))$; and we write '$\mathcal{P}$ is w' to mean $\forall x(\mathcal{P}x \leftrightarrow x \in w)$. ($\mathcal{P}$ is always a one-place predicate symbol.)

Let Ax be a set of $\in$-sentences including $Z_0{}^{(0)}$. We say that an $\in$-formula Φ, having exactly the distinct free variables $x_1, \ldots, x_n$, is *Ax-absolute* if $Ax \vdash$ 'if $\mathcal{P}$ is V, or, for some limit partial universe w, $\mathcal{P}$ is w, and if $\mathcal{P}x_1$ and ... and $\mathcal{P}x_n$, then Φ if and only if $\Phi^{(\mathcal{P})}$.

Metaproposition 1.1. *Any bounded formula is $Z_0^{(0)}$-absolute.*

The reader can see 1.1 in his head.

Of course, when we speak, say, of the formula 'x is a subset of y', we mean the formula $\forall z(z \in x \rightarrow z \in y)$. This style is sloppy but in practice quite clear. In this chapter the statements of the main results are quite precise; but their proofs and the earlier propositions, etc., are occasionally written in a sloppy but intuitively clear way.

Here is a useful example: The formula 'u is a partial universe' means 'for some t, $u \in t$ and t is a tower'. So it is not seen at once to be bounded. However, it is easy to see (in $Z_0^{(0)}$) that if u is a partial universe then u belongs to some tower t which belongs to $P(Pu)$. From this it easily follows that

(2) (a) 'u is a partial universe' is $Z_0^{(0)}$-absolute. (See Problem 3.) Hence, at once, these formulas are also $Z_0^{(0)}$-absolute: 'u is a limit partial universe', 'u is a limit partial universe and not the first such' and 'u is the second limit partial universe'.

We leave as an easy problem (Problem 4) showing that

(3) 'u has class-replacement' is $Z_0^{(0)}$-absolute.

Let the sentence Σ be a universalization of an axiom in $Z_0^{(0)}$.

Theorem 1.2 (In Z_0^*). *Suppose $\mathcal{P}$ is V or for some limit partial universe u, $\mathcal{P}$ is u. Then:*

(a) $\Sigma^{(\mathcal{P})}$;
(b) *if AC then* $(AC)^{(\mathcal{P})}$;
(c) $Reg^{(\mathcal{P})}$.

Proof. As regards (a) and (b), we deal with the two cases of (a) where Σ is the Power Set Axiom or the Separation Axiom; and leave the other (similar) cases to the reader (Problem 5).

Assume the hypothesis of 1.2. Take Σ to be the Power Set Axiom, which can be written

$$\forall A \exists U[(\forall B \in U)(B \subseteq A) \wedge (\forall B \subseteq A)(B \in U)].$$

Thus Σ is $\forall A \exists U \Phi$, where Φ is bounded. To show $\Sigma^{(\mathcal{P})}$, assume that $\mathcal{P}A$. Clearly, $\mathcal{P}$ is 'closed under' taking power sets; so letting $U = P(A)$ we have $\mathcal{P}U$. Obviously $\Phi(A, U)$. Hence by 1.1, $\Phi^{(\mathcal{P})}(A, U)$. So $\Sigma^{(\mathcal{P})}$, as desired.

Secondly, let Σ be a typical case of Separation, namely: $\forall A \exists B \forall x(x \in B \leftrightarrow x \in A \wedge \Psi(x; z_1, \ldots, z_k))$. Suppose $\mathcal{P}A$ and $\mathcal{P}z_1$, etc. By (outside) Separation form $B = \{x \in A : \Psi^{(\mathcal{P})}(x; z, \ldots)\}$. Since $B \subseteq A$, $\mathcal{P}B$. Thus for any x (certainly for any x in $\mathcal{P}$) we have: $x \in B \leftrightarrow x \in A \wedge \Psi^{(\mathcal{P})}$. Hence $\Sigma^{(\mathcal{P})}$ as desired.

As regards (c), suppose $\mathcal{P}x$. We must show that (x belongs to some partial universe)$^{(\mathcal{P})}$. But, from the assumptions about $\mathcal{P}$, clearly there is some partial universe U such that $x \in U$, and U is in $\mathcal{P}$. By (2), (U is a partial universe)$^{(\mathcal{P})}$, finishing the proof of 1.2.

Now let the sentence Σ be a universalization of an $\in$-instance of Replacement.

Theorem 1.3.
(a) *(in Z_0F^*.) If $\mathcal{P}$ is V then $\Sigma^{(\mathcal{P})}$.*
(b) *(in Z_0^*.) If $\mathcal{P}$ is a limit partial universe u and u has class-replacement, then $\Sigma^{(\mathcal{P})}$.*

We may suppose that Σ is a universalization of the $\in$-formula

$$\begin{aligned} &\forall x \exists ! y \Phi(x, y, z_1, \ldots) \rightarrow \\ &\exists B \forall y (y \in B \leftrightarrow \exists x (x \in A \wedge \Phi(x, y, z_1, \ldots)). \end{aligned} \tag{4}$$

Proof of 1.3. (a) We argue in Z_0F^*. Assume $\mathcal{P}$ is V. To show $\Sigma^{(\mathcal{P})}$, assume that $\mathcal{P}A$, $\mathcal{P}z_1$, etc. and that $(\forall x \exists ! y \Phi)^{(\mathcal{P})}$. It follows at once that $\forall x(\mathcal{P}x \rightarrow \exists ! y(\mathcal{P}y \wedge \Phi^{(\mathcal{P})})$. $\mathcal{P}$ is 'transitive' so $\mathcal{P}x$ holds for every $x \in A$. Hence by the Replacement Axiom,

$$\exists ! B \forall y (y \in B \leftrightarrow (\exists x \in A)(\mathcal{P}y \wedge \Phi^{(\mathcal{P})}).$$

Now if we have $\mathcal{P}B$, then it clearly follows that the relativization to $\mathcal{P}$ of the right side of (4) holds, as desired. Obviously, '$B \subseteq \mathcal{P}$'. But, by 8.1.3, any subset of V is in V so $\mathcal{P}B$ and (a) is proved.

(b) Now argue in Z_0^* and assume $\mathcal{P}$ is the limit partial universe u. The proof of (a) is valid as written except that Replacement is not needed to form B since $B \subseteq u$, and that $B \in u$ now follows because u has class-replacement.

Theorem 1.4.

(a) *(In $(Z_0F$ + Infinity$)^*$.) If $\mathcal{P}$ is V then (Infinity)$^{(\mathcal{P})}$.*
(b) *(In Z_0^*). Let u be a limit partial universe, not the first, and suppose $\mathcal{P}$ is u. Then (Infinity)$^{(\mathcal{P})}$.*

Proof. (a) (We work in $(Z_0F$ + Infinity$)^*$.) There is a Peano structure $\underline{A}$. By § 1 and § 2 of Chapter 7 (which were in Z_0F), we can form Ord $\underline{A}$. which is clearly ω, so ω exists. By § 1 of Chapter 8, it is clear that ω and hence $(\omega, Sc, 0)$ are in $V = \mathcal{P}$. Hence, by (1)(b), ($(\omega, Sc, 0)$ is a Peano structure)$^{(\mathcal{P})}$; that is, (Infinity)$^{(\mathcal{P})}$.

(b) (we work in Z_0^*) Let u_0 be the first limit partial universe (which exists since u is the second!). Let A be the set of all partial universes belonging to u_0. Let S be on A with $S(x) = Px$. It is easy to see that $(A, S, \emptyset)$ is a Peano structure and belongs to $u = \mathcal{P}$. Thus, by (1)(b), (Infinity)$^{(\mathcal{P})}$, and the proof of 1.4 is complete.

We will later add to the series 1.1 – 1.4. But we now have more than enough to derive our first independence result. (We will use without mention the fact that, e.g., ZF^* is a conservative extension of $ZF^{(0)}$.(Cf. § 5 of Chapter 9.)

Metatheorem 1.5. (*von Neumann* [*1929*].) *If $ZF^{(0)}$ – Reg is consistent, one can consistently add Reg to it. The same applies to $ZFC^{(0)}$ – Reg, and to $Z_0F^{(0)}$.*

Proof. First we argue in Z_0F^*. Let $\mathcal{P}x$ if and only if Vx. By 1.2 (a), (c) and 1.3(a), we have $\Sigma^{(\mathcal{P})}$ (Σ any sentence in $Z_0F^{(0)}$ + *Reg*). Also, by 1.2(b), if AC then $AC^{(\mathcal{P})}$; and by 1.4(a), if Infinity then Infinity$^{(\mathcal{P})}$.

All three cases of 1.5 now follow at once (using 9.4.5 on interpretations).

Problems

1. Do (1)(a) and (1)(b).
2. Do (1)(c) and (1)(d).
3. Prove (2)(a).
4. Prove (3).
5. Complete the proof of 1.2(a),(b) by dealing with the remaining cases.
6. Improve (2) by showing that '*u* is a partial universe' can even be written in $Z_0^{(0)}$ as a bounded formula. (Hint: go back 2 steps before *u* before considering towers.)

§ 2. Three other consistency results

We show next that *the Axiom of Infinity cannot be proved from the other axioms of ZFC, if they are consistent.* In fact an (apparently) weaker assumption can be made:

Metatheorem 2.1. *If $Z_0F^{(0)}$ is consistent, so is $ZFC^{(0)}$ – Infinity + (~ Infinity) (which is $Z_0F^{(0)}$ + Reg + Choice + ~ (Infinity)).*

Proof. Assume $Z_0F^{(0)}$ is consistent. Then, by Metatheorem 1.5, so is $Z_0F^{(0)}$ + *Reg*. We claim that $Z_0F^{(0)}$ + *Reg* + (~ Γ) is consistent, where Γ is 'There exists a limit partial universe.' If $Z_0F^{(0)}$ + *Reg* ⊢ ~ Γ, this is obvious; so we can assume that T = $Z_0F^{(0)}$ + *Reg* + Γ *is consistent.* Let $\Psi(x)$ be '*x* belongs to the first limit partial universe'. By 1.2(a), (c), $Z_0^{(0)}$ + *Reg is interpreted by* Ψ in T. By 8.2.5, T ⊢ 'V_ω has class replacement'. Hence, by 1.3(b), Replacement is interpreted in T by Ψ. Obviously (using (2)), ~ Γ is so interpreted. Hence (by 9.4.5), T' = $Z_0F^{(0)}$ + *Reg* + (~ Γ) *is consistent.* Theorem 2.1 will be proved if we show that T' ⊢ AC ∧(~ Infinity).

Argue in T'. From Chapter 8, § 1, we obtain at once ~ Infinity. Also, by 8.2.4, every set is finite. Hence, by Problem 14, § 2, Chapter 2 (on the axiom of choice for finitely many sets) we obtain AC, as desired.

Aside. By expending more effort in avoiding Replacement, a patient reader can show that the hypothesis of 2.1 can be weakened to '$Z_0^{(0)}$ is consistent', and that $Z_0^{(0)} + Reg + (\sim \Gamma)$ implies the whole $ZFC^{(0)}$ – Infinity + (~ Infinity).

The next Metatheorem, 2.2, is a close relative of 2.1, but at an exorbitantly high level! For a discussion of 2.2 see the end of § 6 of Chapter 7.

Metatheorem 2.2. *If $ZFC^{(0)}$ is consistent, so is $ZFC^{(0)}$ + 'there is no inaccessible cardinal > $\aleph_0$'.*

Proof. See Problem 1. Hints: Assume $ZFC^{(0)}$ is consistent. Let $\Delta(u)$ be "u is a limit universe, not the first, and u has class replacement." Let Γ be $\exists u\ \Delta(u)$. Show $ZFC^{(0)} + \sim \Gamma$ is consistent. Then show $ZFC^{(0)} + \sim \Gamma \vdash$ 'there is no inaccessible > $\aleph_0$'. (Cf. 8.2.6)

The last metatheorem, 2.3, shows *one cannot derive Replacement from the other axioms of $ZFC^{(0)}$ (assumed consistent), or indeed even a very low-level consequence of Replacement asserting the existence of $V_{\omega+\omega}$.* 2.3 is again a close relative of 2.1 and 2.2.

Metatheorem 2.3. *If $ZC^{(0)}$ is consistent, then one can consistently add to it 'There is no second limit partial universe.' The same applies to $Z^{(0)}$.*

On the other hand, in $Z^{(0)}$ one *can* easily prove (making an essential use of our form of *Reg*) the existence of the *first* limit partial universe, say W, and of PW, PPW, ...

Proof of 2.3. Assume $ZC^{(0)}$ is consistent. Let $\Delta(u)$ be 'u is the second limit partial universe'; and let Γ be $\exists u\ \Delta(u)$. We want to show $ZC^{(0)} + \sim \Gamma$ is consistent. If $ZC^{(0)} \vdash \sim \Gamma$, we are through, so we can assume $T = ZC^{(0)} + \Gamma$ is consistent. Let Ψ be 'x belongs to the second limit partial universe.' Ψ is an interpretation in T of $Z_0^{(0)} + Reg$ + Infinity, by 1.2(a), (c) and 1.4(b), and of Choice, by 1.2(b); using (2) of § 1 it is easy to see that $\sim \Gamma$ is also so interpreted. (Aside: The fact that we carried out most of § 1, through 1.4, in $Z_0^{(0)}$, rather than $Z_0F^{(0)}$, has just been used for the one and only time.) By 9.4.5, 2.3 now follows, for $ZC^{(0)}$. The case of $Z^{(0)}$ is almost identical. Change 'ZC' to 'Z' in the above, and drop the passage about Choice.

Informal remarks. In interpreting one set theory in another, if (as above) $\in$ is not reinterpreted, one can still relativize to other $\mathcal{P}$'s than the V_δ's or V which alone have been used here. By a theorem (of Gödel, Mostowski, and Shepherdson), every such 'inner model' is 'isomorphic' to one in which $\mathcal{P}$ is *transitive.* (Ours are supertransitive.)

The most famous transitive (but not supertransitive) interpretation is the 'constructible sets', by which Gödel, [F 1940], established that $ZFC^{(0)}$ + *GCH* is consistent (if $ZFC^{(0)}$ is).

Paul Cohen in 1963 (cf. [L 1966]) established a number of remarkable independence results. These included the independence of Choice from $ZF^{(0)}$ (if $ZF^{(0)}$ is consistent) and the independence of CH from $ZFC^{(0)}$ (if the latter is consistent). Shepherdson [1953] had shown that inner models could not be used for further independence results! Cohen does not use inner models, literally. His method is so delicate that one cannot decide whether the *word* 'interpretation' should be applied to it or not!

Problem

1. Prove 2.2.

§ 3. Informal note on the set theories *NB* and *M*

The best-known axiomatization of set theory, after $ZFC^{(0)}$, is the theory NB of von Neumann-Bernays as presented in Gödel [F 1940]. The next most used theory is the theory M of A.P. Morse – which was first presented in print by J. L. Kelley [1955] in an appendix to his book on topology.

Speaking roughly, both NB and M differ from $ZFC^{(0)}$ in considering not only *sets*, as in $ZFC^{(0)}$, but also collections of sets (such as the collection of all sets), which are called *classes*. In M we intend to consider *all* possible collections of sets (just as in $ZFC^{(0)}$ our intent is to consider *all* subsets of any given set). However, in NB (as reflected in the choice of axioms) it is our intention to consider as classes only *certain* collections of sets – roughly speaking the "definable" ones. NB is a subtheory of M.

Now we shall speak sloppily (not just roughly!). Let Θ be the first inaccessible cardinal $> \omega$. It is known that $V_{\Theta+1}$ is a 'model' of M (the sets of highest rank acting as classes). Moreover, $\Theta + 1$ is the smallest ordinal α such that V_α is a model of M. On the other hand, V_Θ is similarly a model of ZFC (cf. 8.2.6 and 1.3(b) above); but there are ordinals $\alpha < \Theta$ such that V_α is a model of $ZFC^{(0)}$ cf. Montague-Vaught [1959]). Let α_0 be the smallest such α. Finally, $V_{\Theta+1}$ is a model of NB. But so is $V_{\alpha_o} \cup W$, for some $W \subseteq P(V_{\alpha_o})$. On the other hand, M has no such model for α_o (but does for some $\alpha < \theta$.

From a finitary point of view, it can be shown (though it is now harder, cf. Shoenfield [1954]) that *NB is a conservative extension* of $ZFC^{(1)}$, just as $ZFC^{(1)}$ is of $ZFC^{(0)}$ (cf. Chapter 9, § 5). Consequently, a sentence of the language of $ZFC^{(0)}$ is provable in $ZFC^{(0)}$ if and only if it is provable in NB. On the other hand, in M *many* new such sentences are provable. Thus, NB should be grouped with $ZFC^{(0)}$ and not with M, even though NB and M are "the ones with classes."

Finally, however, we are interested actually not just, say, in $ZFC^{(0)}$, but in the family of "$ZFC^{(0)}$-type set theories" obtained by adding to $ZFC^{(0)}$ further axioms, such as "there exists an inaccessible cardinal $\neq \omega$." It turns out that all three families ($ZFC^{(0)}$-type, NB-type, M-type) are essentially equivalent. For example, NB is stronger than $ZFC^{(0)}$, but $ZFC^{(0)}$ + "there is an inaccessible $\Theta > \omega$" is stronger than NB, since $V_{\Theta+1}$ is a "model" of NB. Thus the differences among all the set theories are in a way very minor.

For precise and full versions of the above, see Levy [1979].

11. MORE ON CARDINALS AND ORDINALS

The two sections of this chapter discuss two unrelated topics. This work depends on Chapters 1-7 but *not* on 8-10. We will work in *ZFC* − *Reg*.

§ 1. Character of cofinality

Let κ be infinite. We define cf κ (*the character of cofinality of* κ) *to be the smallest infinite cardinal* μ *such that* κ *has a cofinal subset of power* μ. By 7.6.5, we can also say: cf κ *is the smallest infinite* μ *such that* κ *can be expressed as a sum of* μ *cardinals each less than* κ. *Obviously* cf $\kappa = \kappa$ *if and only if* κ *is regular.*

We give some examples. Clearly,

$$\text{cf}\,\aleph_\omega = \aleph_0;\ \text{cf}\,\aleph_{\omega_1} = \aleph_1;\ \text{and cf}\,\aleph_{\omega_\omega} = \aleph_0.$$

We can now make a beautiful application of the König-Zermelo Theorem (4.3.2):

Theorem 1.1. $\kappa^{\text{cf}\,\kappa} > \kappa$, *for any infinite* κ.

Proof. Let cf $\kappa = \mu$. Then $\kappa = \Sigma_{\xi < \mu}\lambda_\xi$, where, for each $\xi < \mu$, $\lambda_\xi < \kappa$. Then by König-Zermelo, $\kappa = \Sigma_{\xi < \mu}\lambda_\xi < \Pi_{\xi < \mu}\kappa = \kappa^\mu = \kappa^{\text{cf}\,\kappa}$.

Theorem 1.1 has two remarkable corollaries (1.2 and 1.3 below), each partially answering a very natural question.

We know that if $\kappa = 2^\lambda$ or even μ^λ, where λ is infinite, then $\kappa^{\aleph_0} = (\mu^\lambda)^{\aleph_0} = \mu^{\lambda\aleph_0} = \mu^\lambda = \kappa$. One might suspect that: *for every* $\kappa > \aleph_0$, $\kappa^{\aleph_0} = \kappa$. (For example, assuming *GCH*, this *is* true for $\kappa = \aleph_1, \aleph_2, \cdots, \aleph_n, \cdots$.) But since $\text{cf}(\aleph_\omega) = \aleph_0$, 1.1 implies at once

Corollary 1.2. $\aleph_\omega^{\aleph_0} > \aleph_\omega$.

The same argument obviously yields more, but we shall be content with this example (and also some parts of 1.5 below).

The second application of 1.1 is to the question (*): For what β can $2^{\aleph_\alpha}$ possibly equal $\aleph_\beta$? As we mentioned in § 7 of Chapter 7, very little can be said. But we do have:

Corollary 1.3. *If κ is infinite, then* $\mathrm{cf}(2^\kappa) > \kappa$.

Proof. $(2^\kappa)^{\mathrm{cf}(2^\kappa)} > 2^\kappa$. But $(2^\kappa)^\kappa = 2^\kappa$. So $\mathrm{cf}2^\kappa > \kappa$.

Thus, for example, we do know at least that $2^{\aleph_0} \neq \aleph_\omega$!

Recent work concerning (*) by Silver and others is mentioned in § 7 of Chapter 7.

Mathematicians fairly often assume the Generalized Continuum Hypothesis (just to see what happens), even though not many people have suggested it is (by itself) a natural axiom. (Rather more seem to feel that $2^{\aleph_0} \neq \aleph_1$ is natural!) However, in [F1940] Gödel defined a special kind of sets called *constructible sets* and showed that if we consider only these sets, all the axioms of $ZFC^{(0)}$ hold and also *GCH* holds. The constructible sets are considered to be well worth studying even by people who are convinced they are by no means all sets. This state of affairs has put added weight on the many results known to follow from *GCH*.

All of that leads up to the next theorem, in which we assume *GCH*. We find that we are able to 'calculate' $\aleph_\alpha^{\aleph_\beta}$ in all cases (using 'cf' and applying 1.1).

Theorem 1.4$^{(GCH)}$. *Let κ and λ be infinite.*

(a) *If* $\lambda < \mathrm{cf}\kappa$ *then* $\kappa^\lambda = \kappa$.
(b) *If* $\mathrm{cf}\kappa \leq \lambda \leq \kappa$ *then* $\kappa^\lambda = \kappa^+$.
(c) *If* $\kappa < \lambda$, *then* $\kappa^\lambda = \lambda^+$.

Proof. For (a), assume $\lambda < \mathrm{cf}\kappa$. The key step (which is often used) is this: If $f\colon \lambda \to \kappa$ then (by the definition of cf) $f\colon \lambda \to \alpha$ for some $\alpha < \kappa$. Now the rest of (a) plus (b) and (c) are routine. (See Problem 1.)

In the rest of § 1 we will show (in the form of a long problem) how to define $\mathrm{cf}\underline{A}$ for any order $\underline{A}$.

Theorem 1.5. (a) *Any order $\underline{A}$ has a cofinal subset which is a well-order* (*as a substructure*) *and has ordinal* $\leq \overline{\overline{A}}$.

The proof is Problem 2. *AC* must be used. One can do the construction in two steps (first getting just a well-ordered cofinal subset). (It can also be done in a more tricky way in one step.). The proof is not trivial, but most readers will succeed!

We can now define cf $\underline{A}$ to be the least α which is the ordinal of some cofinal well-ordered subset of $\underline{A}$. If $\underline{A}$ has a last element, cf $\underline{A}$ = 1; henceforth we will assume $\underline{A}$ has no last element. Obviously, if B is a cofinal subset of $\underline{A}$ and C is a cofinal subset of B then C is a cofinal subset of $\underline{A}$. Using this fact and (a) the reader will readily show (Problem 3):

Theorem 1.5. (b) *cf $\underline{A}$ is a regular cardinal (initial ordinal). Also, if κ is an infinite cardinal, the new* $\mathrm{cf}(\kappa, \in_\kappa)$ = *the old* cf κ.

Theorem 1.5. (c) *Let B be a cofinal subset of $\underline{A}$ (and write $\underline{B} = \underline{A}|B$). Then* cf $\underline{A}$ = cf $\underline{B}$.

This proof (Problem 4) requires a new, but natural argument.

Corollary 1.5. (d) $\mathrm{cf}\,\aleph_\alpha = \mathrm{cf}(\alpha, \in_\alpha)$, *if α is a limit ordinal.*

cf $\underline{A}$ can be called the right-hand character of $\underline{A}$. Then the left-hand character is taken as $\mathrm{cf}(A, \geq)$ or perhaps as its * (reverse).

Now any element x of $\underline{A}$ determines two 'characters', the right-hand character of $\underline{A}_x$ and the left-hand character of $\{y\colon y > x\}$. Obviously all of these characters are important 'invariants' for the study of orders.

Problems

1. Prove 1.4(c) and (d).
2. Prove 1.5(a).
3. Prove 1.5(b).
4. Prove 1.5(c).

§ 2. More on ordinal arithmetic

We consider now only ordinals, rather than arbitrary order types as in Chapter 5. First we put the + and · of Chapter 5 together with the ≤ of Chapter 7, § 2. The following are immediate:

Proposition 2.1.

(a) $\alpha < \beta$ *if and only if for some* $\gamma \neq 0$, $\alpha + \gamma = \beta$.

(b) $\alpha \leq \beta$ *if and only if for some* γ, $\alpha + \gamma = \beta$.

Over all order types we could have defined (as in (b)); $\sigma \leq \tau$ if and only if for some ρ, $\sigma + \rho = \tau$. A notion at least as natural is: $\sigma \lesssim \tau$ if and only if σ is the order type of a *substructure* of some order of type τ.

If $\sigma \le \tau$, obviously $\sigma \precsim \tau$. But the reverse fails badly; for example, clearly $\eta \precsim 1 + \eta$ but $\eta \not\le 1 + \eta$. Nevertheless, 7.1.3 stated exactly that for ordinals:

2.1. (c) *$\alpha \le \beta$ if and only if $\alpha \precsim \beta$.*

The next proposition discusses the preservation of $\le$ and $<$:

Proposition 2.2.
(a) *If $\alpha < \beta$ then $\gamma + \alpha < \gamma + \beta$.*
(b) *If $\alpha < \beta$ then $\alpha + \gamma \le \beta + \gamma$ (but not always $<$; for example, clearly, $1 + \omega = 2 + \omega$).*
(c) *If $\alpha < \beta$ and $\gamma \ne 0$, then $\gamma\alpha < \gamma\beta$.*
(d) *If $\alpha < \beta$ and $\gamma \ne 0$ then $\alpha\gamma \le \beta\gamma$ (but not always $<$; for example, $2\omega = 3\omega$).*

Proof. (a) Let $\alpha + \delta = \beta$, with $\delta \ne 0$. The $\gamma + \beta = \gamma + (\alpha + \delta) = (\gamma + \alpha) + \delta$, so $\gamma + \alpha < \gamma + \beta$. (c) is similar. Consider (b). Since $\alpha \le \beta$ we have $\alpha \precsim \beta$. So clearly $\alpha + \gamma \precsim \beta + \gamma$ so $\le$ holds by 2.1. (d) is similar. (See Problem 1.) Note the use of $\precsim$.

By (a), if $\alpha \le \beta$ then there is exactly one γ such that $\alpha + \gamma = \beta$ (but *not* similarly for $\gamma + \alpha$). Thus there is one kind of *subtraction* for ordinals. (We do not discuss it further.)

Trying to imitate both statements and proofs of ordinary number theory as closely as we can, we obtain the Division Algorithm:

Theorem 2.3. *If $\beta \ne 0$ and α are given, there are uniquely determined ordinals ξ and η such that*

$$\alpha = \beta \cdot \xi + \eta \text{ and } \eta < \beta.$$

Proof. The list $\beta \cdot 0, \beta \cdot 1, \cdots, \beta \cdot \xi, \cdots$ is strictly increasing (by 2.2(c)). Hence, by Theorem 1.1 of Chapter 7, always $\beta \cdot \xi \ge \xi$. It obviously follows that for some γ, $\beta \cdot \gamma > \alpha$ and in fact we take γ to be the first such. Next we show γ is not a limit ordinal; then we take $\gamma = \xi + 1$, etc. (See Problem 2.)

Among *ordinals* there is a second way to define $+$ and $\cdot$ (from that of Chapter 5): They can be defined by recursion, using the following conditions (which are clearly valid), where δ is any limit ordinal;

$$\begin{cases} \alpha + 0 = \alpha, \\ \alpha + (\beta + 1) = (\alpha + \beta) + 1, \\ \alpha + \delta = \sup_{\beta < \delta}(\alpha + \delta), \end{cases} \qquad \begin{cases} \alpha \cdot 0 = 0, \\ \alpha \cdot (\beta + 1) = \alpha\beta + \alpha, \\ \alpha \cdot \delta = \sup_{\beta < \delta} \alpha\beta. \end{cases}$$

(The two bottom conditions are new, but very easily verified.)

We now define exponentiation α^β for ordinals recursively by specifying that

$$\begin{cases} \alpha^0 = 1 \\ \alpha^{\beta+1} = \alpha^\beta \cdot \alpha \\ \alpha^\delta = \sup_{\beta<\delta} \alpha^\beta \ (\delta \text{ a limit ordinal}) \end{cases}$$

Aside. There is another way of defining α^β (in something like the style of Chapter 5) but this time it is not as simple as the recursive definition. (The other procedure is sketched in Problem 8.)

More or less as for ordinary numbers we now obtain

Proposition 2.4.

(a) $\alpha^{\beta+\gamma} = \alpha^\beta\alpha^\gamma$.

(b) $(\alpha^\beta)^\gamma = \alpha^{\beta\gamma}$.
(Since multiplication is not commutative, $(\alpha\beta)^\gamma$ is rarely $\alpha^\gamma\beta^\gamma$.)

(c) *If $\alpha < \beta$ then $\alpha^\gamma \le \beta^\gamma$ (but not always $<$).*

(d) *$0^\gamma = 0$ if $\gamma \ne 0$, and $1^\gamma = 1$.*

(e) *$\alpha^1 = \alpha$, $\alpha^2 = \alpha\alpha$; for natural numbers m, n, the new (ordinal) m^n = the old (cardinal) m^n.*

(f) *If $\beta < \gamma$ and $\alpha \ge 2$ then $\alpha^\beta < \alpha^\gamma$.*

Proof. (a)–(d) are all proved using induction (see Problem 3). All of (e) is obvious. (f) can be proved without induction (Problem 4).

For ordinals (in fact even for order types) we had $\overline{\overline{\alpha + \beta}} = \overline{\overline{\alpha}} + \overline{\overline{\beta}}$ (the second + is cardinal) and similarly, $\overline{\overline{\alpha \cdot \beta}} = \overline{\overline{\alpha}} \cdot \overline{\overline{\beta}}$. *The analogue for exponentiation is completely false!* For example, $\overline{\overline{2^\omega}} = \aleph_0$ (but $2^{\overline{\overline{\omega}}} = 2^{\aleph_0}$). In fact, the following holds in general:

Theorem 2.5^{AC}. If $\alpha > 1$, $\beta > 0$, and at least one of α and β is infinite, then $\overline{\overline{\alpha^\beta}} = \max(\overline{\overline{\alpha}}, \overline{\overline{\beta}})$.

For the proof (of moderate difficulty) see Problem 5. Theorem 2.5 will not be used below.

Perhaps surprisingly, ordinal arithmetic *is* occasionally applied in other parts of mathematics, often to construct examples. The last topic in this section will be the representation of ordinals to any 'base' ≥ 2. (The base ω is often used.) This representation is one of the nicest parts of ordinal arithmetic and is also one of the things most often used in applications. Naturally, it is due to G. Cantor!

Theorem 2.6. *Suppose $\beta \ge 2$.*

(a) Let $\zeta \ne 0$. *Then there are α, ρ, η such that $0 < \eta < \beta$, $\rho < \beta^\alpha$, and $\zeta = \beta^\alpha \cdot \eta + \rho$; and this representation is unique.*

Proof. By 2.4 (f), the ordinals $\beta^0, \beta^1, \cdots, \beta^\xi, \cdots$ are strictly increasing. So (by 7.1.1) $\beta^\xi \geq \xi$, for all ξ. Hence we can take γ to be the least ordinal such that $\beta^\gamma > \zeta$. Clearly $\gamma \neq 0$, since $\zeta \neq 0$. If γ were a limit ordinal, then $\beta^\gamma = \sup_{\nu < \gamma} \beta^\nu$, so clearly, for some $\nu < \gamma, \beta^\nu > \zeta$, a contradiction. Thus we can write γ as $\alpha + 1$. Thus $\beta^\alpha \leq \zeta < \beta^{\alpha+1}$. By 2.3, we can write $\zeta = \beta^\alpha \cdot \eta + \rho$ where $\rho < \beta^\alpha$. Now $\eta \neq 0$ (else $\zeta < \beta^\alpha$) and $\eta < \beta$ (else $\zeta \geq \beta^{\alpha+1}$). So we have the desired representation.

For uniqueness, notice that if α, ρ, and η are as in (a), then $\beta^\alpha \leq \zeta < \beta^{\alpha+1}$ (the second inequality holds because $\zeta = \beta^\alpha \cdot \eta + \rho < \beta^\alpha \cdot \eta + \beta^\alpha = \beta^\alpha(\eta + 1) \leq \beta^\alpha\beta = \beta^{\alpha+1}$). Thus α is determined. It follows now from 2.3 that η and ρ are also determined.

(b) *Suppose* $\alpha_0 > \alpha_1 > \cdots > \alpha_{n-1}$, *and* $0 < \eta_0, \eta_1, \ldots, \eta_{n-1} < \beta$. *Then* $\beta^{\alpha_0} \cdot \eta_0 + \beta^{\alpha_1} \cdot \eta_1 + \cdots + \beta^{\alpha_{n-1}} \cdot \eta_{n-1} < \beta^{\alpha_0 + 1}$.

Proof. This is clear for $n = 0$. Suppose it holds in general for k, and consider the case $n = k + 1$. By the induction hypothesis, the left-hand side (in (b)) is equal to

$$\beta^{\alpha_0} \cdot \eta_0 + (\beta^{\alpha_1} \cdot \eta_1 + \cdots) < \beta^{\alpha_0} \cdot \eta_0 + \beta^{\alpha_1 + 1} \leq$$
$$\beta^{\alpha_0} \cdot \eta_0 + \beta^{\alpha_0} = \beta^{\alpha_0}(\eta_0 + 1) \leq \beta^{\alpha_0 + 1}.$$

Thus (b) holds for $k + 1$, and (b) is proved.

Theorem 2.6. (c) *(Representation to the base $\beta \geq 2$). For any ζ there exist* n; $\alpha_0, \ldots, \alpha_{n-1}$; *and* $\beta_0, \ldots, \beta_{n-1}$ *such that* $\alpha_0 > \alpha_1 > \cdots > \alpha_{n-1}$, *and* $\beta_i \neq 0$ *for all* $i < n$, *and*

$$\zeta = \beta^{\alpha_0} \cdot \eta_0 + \beta^{\alpha_1} \cdot \eta_1 + \cdots + \beta^{\alpha_{n-1}} \cdot \eta_{n-1};$$

moreover, the representation is unique.

Proof. (See Problem 6.) (c) follows fairly easily from (a) and (b).

Note. The finite sums above (in (b) and (c)) *can* be understood as the ordered Σ of Chapter 5. But it is simpler to think of them as coming from the binary + by an ordinary recursion, as, e.g., in elementary group theory.

The remainder of §2 can be considered optional.

Obviously $\rho \neq 0$ is the type of a non-empty final segment of α (i.e., $(\alpha, \in_\alpha)$) if and only if $\alpha = \gamma + \rho$ for some γ. For certain $\alpha \neq 0$, every non-empty final segment of α has type α. Applying 2.6, one can prove

Theorem 2.7. *The following are equivalent for* $\alpha \neq 0$: (a) *Every non-empty final segment of* α *has ordinal* α, (b) $\alpha = \omega^\delta$ *for some* δ. (Problem 7 – of moderate difficulty.)

Except for 2.5, AC was never used in this section, except that, in our development, AC was used in § 4 of Chapter 7 to define $\overline{\overline{A}}$ and Tp$\underline{A}$ and then reabsorb Chapters 2-5. We have only needed Chapter 5 and we have needed Tp$\underline{A}$ only for well-ordered $\underline{A}$. But for such $\underline{A}$ we can use the Ord $\underline{A}$ of § 2, Chapter 7 as Tp$\underline{A}$. Thus AC did not really have to be used.

One can also eliminate Replacement. On the face of it, the recursive definition of α^β leans heavily on Replacement. We shall not try to eliminate Replacement here. But one way of doing it involves something of considerable independent interest which we shall discuss – namely a way of dealing with α^β in a manner parallel to the original non-recursive definitions of $\alpha + \beta$ and $\alpha \cdot \beta$:

Let $K_{\alpha,\beta}$ be the set of all $f: \beta \to \alpha$ such that $f(\gamma) = 0$ for all but finitely many $\gamma < \beta$. Given $f, g \in K_{\alpha,\beta}$ we put $f < g$ if among the finitely many γ for which $f(\gamma) \neq g(\gamma)$ the *largest* such γ has $f(\gamma) < g(\gamma)$. Now it can be shown (Problem 8 – of moderate difficulty) that $\leq$ well-orders $K_{\alpha,\beta}$ in the type which is α^β. (Working without Replacement, this type can be taken as the *definition* of α^β.)

Problems

1. Prove 2.2.
2. Complete the proof of 2.3.
3. Prove 2.4 (a)-(d).
4. Prove 2.4 (f).
5. Prove 2.5.
6. Prove 2.6 (c).
7. Prove 2.7.
8. Show $(K_{\alpha,\beta}, \leq)$ is a well-order, and its type is α^β.

APPENDIX

Proofs of some results in Chapter 9

Proposition 3.6. $\vdash \Phi \leftrightarrow \Phi'$ *for any variant* Φ' *of* Φ.

Proof. First a lemma:

Lemma. *Suppose y does not occur in* Φ. *Then: (a) y is free in* $\Phi\binom{x}{y}$ *at exactly the places where x is free in* Φ. *(b)* $\vdash \forall x\Phi \leftrightarrow \forall y\Phi\binom{x}{y}$.

(a) is obvious. We show $\leftarrow$ in (b) ($\rightarrow$ is similar but easier). Since y is not in Φ, the hypotheses of 3.5 hold, and 3.5 gives $\vdash \forall y\Phi\binom{x}{y} \rightarrow \Phi\binom{x}{y}\binom{y}{x}$. One now easily gets $\vdash \forall y\Phi\binom{x}{y} \rightarrow \forall x\Phi$, using Axioms II and III (since x is not free in $\forall y\Phi\binom{x}{y}$). So the Lemma is proved.

Let us say that Φ' is a *special* variant of Φ if it is a variant and no bound variable of Φ' occurs (at all) in Φ. It is easy to see (using 'new' variables) that *any two variants* have a common special variant. So 3.6 reduces to showing (*) *each* Φ *is equivalent to each of its special variants.*

The case where Φ is atomic is trivial. If Φ is $\Phi_1 \rightarrow \Phi_2$ clearly any special variant Ψ of Φ is of the form $\Psi_1 \rightarrow \Psi_2$ where Ψ_i is a special variant of Φ_i for $i = 1, 2$. By the induction hypothesis, $\Phi_1 \leftrightarrow \Psi_1$ and $\Phi_2 \leftrightarrow \Psi_2$ so clearly $\Phi \leftrightarrow \Psi$. The case where Φ is $\sim\Phi_1$ is similar. Now let Φ be $\forall x\Phi_1$, and let Ψ be any special variant of Φ. Ψ can be written $\forall y\Psi_1$, where x and y are different. Clearly, (**) Φ_1 and Ψ_1 are variants except that Φ has free occurrences of x exactly where Ψ has free occurrences of y. Now Ψ_1 has no bound occurrence of x (since Ψ is a special variant of Φ); and also x is not free in Ψ_1 (as then it would be free in Ψ and hence in $\Phi = \forall x\Phi_1$). So we can apply the Lemma to get $\vdash \forall y\Psi_1 \leftrightarrow \forall x\Psi_1\binom{y}{x}$. But, by (**), $\Psi_1\binom{y}{x}$ is a variant, and so clearly a special variant, of Φ_1. Thus, by the induction hypothesis, $\vdash \Phi_1 \leftrightarrow \Psi_1\binom{y}{x}$. Hence $\vdash \forall x\Phi_1 \leftrightarrow \forall y\Psi_1$, i.e., $\vdash \Phi \leftrightarrow \Psi$. Thus (*) (and 3.6) are proved.

Theorem 4.1. (a) *Assume S is safe for Φ and $\vdash \Phi$. Then $\vdash \Phi(S)$.* (*b*) *If $\vdash \Phi$ then $\vdash^{\mathcal{L}'} \Phi$ in any language $\mathcal{L}'$ containing Φ.*

Before beginning the proof, we make two remarks. It is easy to check that

(1) If S is safe for Φ and z is free (or, respectively, bound) in $\Phi(S)$ then either z is free (or bound, respectively) in Φ or else z is a variable of S.

Secondly, it will be convenient (though it is not essential) to consider in the proofs of 4.1 and 5.1 permutations of the set *Vbl.* of all variables. (All our permutations f can have $f(x) = x$ for all but finitely many variables x, so they are 'finitary objects.')

If f is a permutation of *Vbl*, f induces in the obvious way a permutation $\bar{f}$ of the collection of all expressions and even sets of expression, etc. Sometimes we write $\Gamma^{(f)}$ instead of $\bar{f}(\Gamma)$. In defining 'formula', 'axiom of logic', etc., etc., we never singled out any particular variables. Hence $\bar{f}$ 'goes through everything'. For example, if $\Phi_1, \ldots, \Phi_n$ is a proof of Φ then $\Phi_1{}^{(f)}, \ldots, \Phi_n{}^{(f)}$ is a proof of $\Phi^{(f)}$. On the other hand, if Σ is any *sentence*, clearly $\Sigma^{(f)}$ is a variant of Σ, and hence $\vdash \Sigma \leftrightarrow \Pi^{(f)}$. In general, for any set Ax of sentences, $Ax^{(f)}$ is the collection of all $\Sigma^{(f)}$ such that Σ belongs to Ax; so clearly, *Ax and $Ax^{(f)}$ are 'equivalent' (have the same theorems).*

We now turn to the

Proof of Theorem 4.1. Assume S is safe for Φ and $\vdash \Phi$. Let $\Phi_1', \ldots, \Phi_p'$ be a proof of Φ. We can obviously construct a permutation f of *Vbl* which takes each variable in Φ to itself and takes every other variable in any Φ_i' into one not in S. Since S was safe for Φ clearly S is safe for each $\Phi_i'^{(f)}$. By the remark above, $\Phi_1'^{(f)}, \ldots, \Phi_p'^{(f)}$ is a proof of $\Phi_p'^{(f)} = \Phi$. Taking $\Phi_i = \Phi_i'^{(f)}$ we have shown:

(2) Φ has a proof $\Phi_1, \ldots, \Phi_p$ such that S is safe for each Φ_i.

We now show:

(3) If $\Phi_1, \ldots, \Phi_p$ is a proof of Φ, S is safe for each Φ_i, and $\mathcal{L}'$ is a language containing all the formulas $\Phi_1(S), \ldots, \Phi_n(S)$, then $\vdash^{\mathcal{L}'} \Phi_k(S)$ for $k = 1, \ldots, p$.

(Thus, in particular, we will have $\vdash \Phi(S)$, proving (a).)

We prove that $\vdash^{\mathcal{L}'} \Phi_k(S)$ by induction on $k < p$. Suppose it is true for all $k' < k$. First suppose that Φ_k is obtained by Modus Ponens from earlier Φ_i's or is a case of Axioms I, II, or IV(a). Then the very same applies to $\Phi_k(S)$ at once since $-(S)$ 'goes through everything'. Hence, clearly, $\vdash^{\mathcal{L}'} \Phi_k(S)$. Next, suppose Φ_k is from Axiom III, say, Φ_k is $\Gamma \to \forall x \Gamma$ where x is not free in Γ. Then $\Phi_k(S)$ is $\Gamma(S) \to \forall x \Gamma(S)$. Now, S is clearly safe for Γ so by (1) if x is free in $\Gamma(S)$, then x is free in Γ (out, by hypothesis), or is a variable of S (impossible since

S is safe for Φ_k). So x is not free in $\Gamma(S)$ and $\Phi_k(S)$ (as above) is a case of Axiom III. Finally, suppose Φ_k is a case of IV(b). To simplify the notation, say (as an example) that Φ_k is $x = y \rightarrow (Qxyxz \leftrightarrow Qxyyz)$ (x, y, z distinct). Then $\Phi_k(S)$ is either Φ_k (so provable in $\mathcal{L}'$) or else Q is a P_i (of S) and so $\Phi_k(S)$ is

$$x = y \rightarrow \left(\Theta_i\binom{x_{i1}\cdots x_{i4}}{xyxz} \leftrightarrow \Theta_i\binom{x_{i1}\cdots x_{i4}}{xyyz}\right)$$

(No collisions occur here and hence this formula is a case of Proposition 3.4 since $\Theta_i(\dot{x}\dot{y}\dot{y}\dot{z})$ is obtained from $\Theta_i(\dot{x}\dot{y}\dot{x}\dot{z})$ by replacing a number of occurrences of x by *free* occurrences of y (this is so because S is safe for Φ). Hence, by 3.4, $\vdash^{\mathcal{L}'}\Phi_k(S)$, and the proof of (3) is complete.

It remains to prove (b). Let $\Phi_1, \ldots, \Phi_p$ be any proof of Φ. Let $\mathcal{P}_1, \ldots, \mathcal{P}_m$ be a list without repetitions of all $\mathcal{Q}$ such that $\mathcal{Q}$ is in some Φ_i but not in Φ. Let γ be a fixed sentence involving only $\approx$ and containing no variables from any Φ_i. For $j \leq m$, let S take $\mathcal{P}_j$ (which is, say, l_j-ary) to γ (or, strictly speaking, to $(\gamma; w_1^j, \ldots, w_{l_j}^j)$ – where the w's are distinct variables not in any Φ_i.) Obviously S is safe for each Φ_i; and each $\Phi_i(S)$ is in the smallest language $\mathcal{L}'$ containing Φ. Hence, by (3), $\vdash^{\mathcal{L}'}\Phi(S)$. But $\Phi(S) = \Phi$, so $\vdash^{\mathcal{L}'}\Phi(S)$, as desired. The proof of 4.1 is complete.

We now turn to the proof of

Metatheorem 5.1. (a) *Suppose* $\mathrm{Z}^{(1)}\vdash \Phi$ *and* ψ *is an instance of* Φ. *Then* $\mathrm{Z}^{(1)}\vdash \psi$. (b) $\mathrm{Z}^{(1)}$ *is a conservative extension of* $\mathrm{Z}^{(0)}$, *i.e., if* Φ *is in* $\mathcal{L}^{(0)}$ *and* $\mathrm{Z}^{(1)}\vdash \Phi$, *then* $\mathrm{Z}^{(0)}\vdash \Phi$.

Proof of 5.1. Any Φ can be written as $\Phi(S)$ with S empty; so (b) will follow if we can prove (a) and

(a′) Moreover, in (a), if Ψ is in $\mathcal{L}_0$ then $\mathrm{Z}^{(0)}\vdash \Psi$.

Assume $\mathrm{Z}^{(1)}\vdash \Phi$ and Ψ is an instance of Φ. Then $\Psi = \Phi(S)$, where

(5)(a) Φ and S are as in (1) of Chapter 9,§ 5.

By the Deduction Theorem, we also have:

(5)(b) $\vdash \Sigma \rightarrow \Phi$, where Σ is a conjunction of sentences which are universalizations of formulas $\Gamma_j(S_j)$, where (for each j) (5)(a) holds for Γ_j and S_j, and Γ_j belongs to Z.

At first, we shall suppose also that

(6) No variable of S is in Σ or any Γ_j or S_j.

(At the end we will see that the general case reduces to this one.)

For the sake of (a′) we extend S to S^* as follows: S^* is defined for $\mathcal{P}_1, \cdots, \mathcal{P}_m, \mathcal{P}_{m+1}, \cdots, \mathcal{P}_n$ where $\mathcal{P}_{m+1}, \cdots, \mathcal{P}_n$ is a list of those $\mathcal{Q}$ occurring in Σ but for which S was not defined. Now (as in the proof of 4.1 (b)), S^* takes each such new $\mathcal{Q}$ (i.e., for which S was not defined) to $(\gamma; \cdots)$ where γ involves only $\approx$ and all the variables of γ are new. Since (5)(a) hold for S, $\Psi = \Phi(S) = \Phi(S^*)$. By (6), S and so also S^* are clearly safe for $\Sigma \rightarrow \Phi$. Hence, by Theorem 4.1, $\vdash \Sigma(S^*) \rightarrow \Phi(S^*)$. By our assumptions and the way S^* was constructed we see that: *If* Ψ *is in* $\mathcal{L}_0$ *then so is* $\Sigma(S^*)$. Hence both (a) and (a′) reduce to showing that $Z^{(1)} \vdash \Sigma(S^*)$, or that for an arbitrary, but fixed j we have:

(7) $$Z^{(1)} \vdash \Gamma_j(S_j)(S^*).$$

(7) will follow at once if we can show that for some S', safe for Γ_j,

(8) $$\Gamma_j(S_j)(S^*) = \Gamma_j(S_j').$$

Just for notation, let

$$S_j = \begin{bmatrix} \mathcal{P}_1^j & \cdots & \mathcal{P}_{m_j}^j \\ (\Theta_i^j;\, x_{11}^j, x_{12}^j, \ldots) & \cdots & \cdots \end{bmatrix}.$$

By (6), S^* is safe for each Θ_i^j. Put

$$S_j' = \begin{bmatrix} \cdots & \mathcal{P}_i^j & \cdots \\ \cdots & (\Theta_1^j(S^*), x_{i1}^j, x_{i2}^j, \ldots) & \cdots \end{bmatrix} i = 1, \ldots, m_j.$$

We easily see that S_j' is safe for Γ_j. (Indeed, if x is in S_j' then (by (1)) either x is in S_j – so not in Γ_j as S_j is safe for Γ_j, or else x is in S^* – so not in Γ_j by (6).) Hence, it remains only to prove (8). One can pretty much see (8) in his head; but a careful proof is nearly a page long. See Problem 1, just after the proof of 5.1.

Thus we have obtained the conjunction of (a) and (a′), assuming (6).

We now return to our position just after (5) (before we assumed (6)). Obviously, there is a permutation f which takes each variable in Φ to itself and takes every other variable in Σ or any Γ_j or S_j to one not in S. (Since S is safe for Φ, $f(x)$ is also not in S if x is in Φ.) The statements (5) and (6) are about $\Sigma, \cdots \Gamma_j \cdots, \cdots S_j \cdots, Z, \Phi$ and S. We claim that

(9) $$\begin{cases} \text{(5) and (6) hold for } \Sigma_1^{(f)}, \cdots, \Sigma_j^{(f)}, \cdots, Z^{(f)}, \Phi \text{ and } S \\ \text{(the last two unchanged).} \end{cases}$$

We have just chosen f so that (6) holds for $\Sigma^{(f)}$, etc., (as in (9)). We know (5) holds for $\Sigma, \ldots, \Gamma_j, \ldots, S_j, \ldots Z, \Phi$, and S. Trivially, (5)(a) holds as in (9), since Φ and S are unchanged. (5)(b) passes over under f (see the remark preceding the proof of 4.1 above) to $\Sigma_1^{(f)}, \ldots, \Gamma_j^{(f)}, \ldots, S_j^{(f)}, \ldots Z^{(f)}$, and $\Phi^{(f)}$. But $\Phi^{(f)} = \Phi$ so (9) holds.

We have proved above that (5) and (6) imply the conjunction of (a) and (a′). Applying this to $\Sigma^{(f)}$, etc., as in (9), we obtain: $Z^{(f)(1)} \vdash \Psi$, and if Ψ is in $\mathcal{L}^{(0)}$, then $Z^{(f)(0)} \vdash \Psi$. But $Z^{(f)(1)} = Z^{(1)(f)}$ which (see above) is equivalent to $Z^{(1)}$. So $Z^{(1)} \vdash \Psi$, as desired. Also $Z^{(f)(0)} = Z^{(0)(f)}$ which is equivalent to $Z^{(0)}$. Thus, if Ψ is in $\mathcal{L}^0$, then $Z^{(0)} \vdash \Psi$. The proof of 5.1 (on the putting back of substitutions) is complete.

Problem 1. Prove (8) as follows: First reduce to showing $\Delta(S_j)(S^*) = \Delta(S_j')$ for Δ an atomic subformula of Γ_j, say Δ is $\mathcal{P}_i^j z_1 z_2 \ldots$. This amounts to

$$(8') \qquad \Theta_i^j(S^*)(s) = \Theta_i^j(s)(S^*) \text{ where } s \text{ is } \left(\begin{smallmatrix} x_{i1}^j & \cdots \\ z_1 & \cdots \end{smallmatrix}\right).$$

(8′) reduces to the case

$$(8'') \qquad \alpha(S^*)(s) = \alpha(s)(S^*) \text{ where } \alpha \text{ is an atomic subformula of } \Theta_i^j \text{ and indeed one of the form } \mathcal{P}_l u_1 u_2 \ldots.$$

One can now obtain (8″) – using the fact that no u_k or x_{il}^j is in Θ.

BIBLIOGRAPHY

When possible, we depend on the bibliographies in the two books Fraenkel [1961] and Levy [1979] (see below). A reference like 'Gödel [F 1940]' directs the reader to the entry 'Gödel [1940]' in the Bibliography of Fraenkel [1961]. If 'L' is used instead of 'F', the reference is similarly to the Bibliography of Levy [1979].

Other References

1936. CHURCH, A., A note on the Entscheidungsproblem. *Journal of Symbolic Logic*, vol. 1, pp. 40-41. Correction, ibid., pp. 101-102.

1961. FRAENKEL, A., *Abstract Set Theory*. Second edition. North-Holland, Amsterdam.

1879. FREGE, G., *Begriffschrift, eine der arithmetischen nachgebildete Formelsprache des reinen Denkens*. Halle.

1930. GÖDEL, K., Die Vollständigkeit der Axiome des logischen Funktionenkalkuls. *Monatshefte fur Math. u. Physik*, vol. 37, pp. 349-360.

1931. GÖDEL, K., Uber formal unentscheidbare Sätze der Principia Mathematica und verwandter Systeme I. *Ibid.*, vol. 38, pp. 173-198.

1960. HALMOS, P., *Naive Set Theory*. D. Van Nostrand, Princeton.

1928. HILBERT, D. and ACKERMAN, W., *Theoretische Logik*. Julius Springer, Berlin.

1978. JECH, T., *Set Theory*. Academic Press, New York.

1956. KALISH, D. and MONTAGUE, R., A simplification of Tarski's formulation of the predicate calculus. *Bulletin Amer. Math. Soc.*, vol. 62, p. 261.

1955. KELLEY, J. L., *General Topology*. Van Nostrand, Princeton.

1979. LEVY, A., *Basic Set Theory*. Springer-Verlag, Berlin, Heidelberg, New York.

1959. MONTAGUE, R. and VAUGHT, R. L., Natural models of set theories. *Fundamenta Mathematica*, vol. 47, pp. 219-242.

1982. MOORE, GREGORY, *Zermelo's Axiom of Choice – its Origin, Development and Influence*. Springer-Verlag, New York.

1929. VON NEUMANN, J., Über eine Widerspruchsfreiheitsfrage in der axiomatischen Mengenlehre. *J. reine angew. Math.*, vol. 166, pp. 227-241.

1953. SHEPHERDSON, J. C., Inner models for set theory, Part III. *Journal of Symbolic Logic*, vol. 18, pp. 145-167.

1954. SHOENFIELD, J., A relative consistency proof. *J. Symbolic Logic*, vol. 19, pp. 21-28.
1936. TARSKI, ALFRED, Der Wahrheitsbegreff in den formalisierten Sprachen. *Studia Philosophica*, vol. 1, pp. 261-405.
1951. TARSKI, A., Remarks on the formalization of the predicate calculus. *Bulletin Amer. Math. Soc.*, vol. 57, p. 81.

Recommendations for more advanced reading.

For further study in straight set theory, we recommend Levy [1979].

For reading in 'meta-set theory' (about the work of Gödel, Cohen, etc.), see Jech [1978]. Also, for an informal but readable outline, see Cohen [L1966].

INDEX

Solutions for Selected Problems

Note: Some problems are not discussed. For example, if a section contains three similar problems, only one may be solved, so the student can still solve the others for himself.

CHAPTER 1

§1. Problem 1. $A \cup (B \cap C) = (A \cup B) \cap (A \cup C)$

Proof. Let $x \in A \cup (B \cap C)$. Then $x \in A$ or $x \in B \cap C$. If $x \in A$, then $x \in A \cup B$ and $x \in A \cup C$ so $x \in$ right side. If $x \in B \cap C$, then $x \in B$ and $x \in C$, so $x \in A \cup B$ and $x \in A \cup C$, so $x \in$ right side. So in either case $x \in$ right side.

Then we have proved left $\subseteq$ right. Now suppose $x \in$ right side, so $x \in A \cup B$ and $x \in A \cup C$. If $x \in A$, then clearly $x \in$ left side. If not, $x \in B$ and $x \in C$ so clearly again $x \in$ left side. Then $x \in$ left side.

Thus we have proved right $\subseteq$ left, so (by above) right = left.

Problem 2. Prove only the first equation.

Problem 5. Assume hypothesis and $x \in A$ (as in 'Note'). Then $x \in A \cup C = B \cup C$. So $x \in B$ (as desired) or $x \in C$ (which we assume). Hence $x \in A \cap C = B \cap C$, so $x \in B$ as desired. So we have proved $A \subseteq B$. One can 'repeat' for $B \subseteq A$. Or having proved that for *any* A, B, C, if $A \cap C = B \cap C$ and $A \cup C = B \cup C$, then $A \subseteq B$ apply this to B, A, C (for A, B, C) getting $B \subseteq A$ at once.

Problem 6. (3), (3), (1), (2), (1), (2).

§2. Problem 1. *Additional hint:* Take $t = \{x : x \in C \text{ and } ?\}$.

§3. Problem 1. Let $x \in$ left side. Then $x \notin \bigcup_{i \in I} A_i$, so for any $i \in I$, $x \notin A_i$, i.e., $x \in \tilde{A}_i$. So $x \in$ right side. Now let $x \in$ right side; the argument is similar but different.

Problem 2. *Note*, the second equation in 3.1(b) should read:

$$B \cup \bigcap_{i \in I} A_i = \bigcap_{i \in I} (B \cup A_i).$$

Problem 4. $\circ$ means bound, $\square$ free.

(a) for all y , $y < x$

(c) $$\sum_{i=1}^{n} (i^2 + 1)$$

§4. Problem 1 Let $\{\{a\}, \{a, b\}\} = \{\{c\}, \{c, d\}\}$. Since $\{a\} \in$ left side, $\{a\} \in$ right side, and hence either $\{a\} = \{c\}$ (case 1) or $\{a\} = \{c, d\}$ (case 2). Continue in this style.

Problem 2. Part on 4.2(c) Both sides are clearly sets of ordered couples. So we can begin with: Let $(x, y) \in A \times (B - C)$. Then $x \in A$ and $y \in B - C$, i.e. $y \in B$ and $y \notin C$. Hence $(x, y) \in A \times B$ and $(x, y) \notin A \times C$, so $(x, y) \in$ right side. Now let $(x, y) \in$ right side...

§5. Problem 1. For example: $G = M/P$ where $P = M \cup F$.

Problem 2. Both sides are sets of ordered couples. Let $(x, y) \in R/S$. Then $(y, x) \in R/S$. So we can choose z such that $(y, z) \in R$ and $(z, x) \in S$. hence $(x, z) \in \breve{S}$ and $(z, y) \in \breve{R}$. But then $(x, y) \in \breve{S}/\breve{R}$. Now let $(x, y) \in$ right side, etc.

Problem 3. Let $y \in$ left side. Then for some x, xRy and $x \in \bigcup_{i \in I} A_i$, i.e. we can choose i such that $x \in A_i$. But then $y \in R[A_i]$ so $y \in$ right side. Other direction 'similar but different'.

Problem 4. Let $x \in$ left side. Then $x \in \operatorname{dom} f$ and $f(x) \in A - B$, etc. etc.

§6. Problem 2 (6.22.) Let $x \in \bigcup_{f \in \prod_{i \in I} J_i} \bigcap_{i \in I} A_{if(i)}$. Then we can choose $f \in \prod_{i \in I} J_i$ such that $x \in \bigcap_{i \in I} A_{if(i)}$, i.e., for all $i \in I$, $x \in A_{if(i)}$. But $f(i) \in J_i$. So for all $i \in I$ there exists $j \in J_i$ (namely $j = f(i)$) such that $x \in A_{ij}$.

§7. Problem 1. 5.4(a)(i) becomes $\underline{A} \underset{\text{id}_A}{\cong} \underline{A}$, etc. etc.

Problem 2. Left implies right: Put $\underline{B} = \underline{A}' \mid f[A]$. Clearly $\underline{A} \underset{f}{\cong} \underline{B}$. The other direction follows at once from the fact that if $\mu, \nu \in f[A]$ then $\mu R' \nu$ iff $\mu(R' \cap (B \times B))\nu$.

Problem 3. *Hint.* For 'at most one' assume R' and R'' both work and show $R = R'$. For 'at least one', let $R' = \{(f(x), f(y)) : xRy\}$ and show R' works.

§8. Problem 1. Further hint: Define F on A by requiring $F(a) = \{b : b \subseteq a\}$ (for each $a \in A$).

Problem 2. First we prove 8.2: We only need to show f is one-to-one and xRy whenever $f(x)R'f(y)$. Write $\underline{A}$ for (A, R) and $\underline{A}'$ for (A', R'). Let $x, y \in A$ and $x \neq y$. Then, since $\underline{A}$ is connected, either xRy or yRx. By symmetry it is enough to deal with the case xRy: By hypothesis $f(x)R'f(y)$. Since $\underline{A}'$ is irreflexive, this implies $f(x) \neq f(y)$ as desired. Now suppose $f(x)R'f(y)$. We want xRy. If not, then $x = y$ or yRx (as $\underline{A}$ is connected). If $x = y$ then $f(x) = f(y)$, contrary to $\underline{A}'$ being irreflexive. If yRx then $f(y)R'f(x)$; but also $(f(x)R'f(y))$, contrary to $\underline{A}'$ being asymmetric. So 8.2 is proved. We have only needed: $\underline{A}$ is connected, and $\underline{A}'$ is asymmetric (as asymmetric implies irreflexive).

Problem 3. Good little problem, but no need to give answer.

Problem 4. Use the complement of the proper initial segment, etc.

Problem 6. (a) $\to$ (b) Outline: Suppose (a) holds and $X \subseteq A$, $x_0 \in X$, and b_0 is a lower bound for X. If $b_0 \in X$, then b_0 is 'clearly' the least element of X which is always a g.l.b. for X. So we can let $b_0 \notin X$. Put $X' = \{y : (\exists x \in X)(x \leq y)\}$. 'Clearly' $b_o \notin X'$. So $\widetilde{X}'$ is non-empty and

bounded above. Hence by (a), $\widetilde{X}'$ has a l.u.b. u. One now shows that u is also a g.l.b. for X (as desired). The rest of Problem 6 must still be proved.

Problem 7. We give the answers, but not the proofs: False, false, true, false.

CHAPTER 2

§1. Problem 1. By 1.2.3 get t as in the Hint. Take $Z = \{t\} \times Y$.

Problem 2. We assume here the equivalence of (a)-(d). We show (b) → (e). Let $\overline{\overline{A'}} = \kappa$. Take A, B as in (b). By the Exchange Principle obtain $C)(A'$ such that $C \sim B - A$. Clearly $B' = A' \cup C$ is as desired.

Problem 3. Suppose $\kappa \leq \lambda$. Then, by 1.5, we can fix A, B with $A \subseteq B$, $\overline{\overline{A}} = \kappa$, $\overline{\overline{B}} = \lambda$. If $\kappa = 0$, we are done.d If not, take $a_0 \in A$. Let f be on B with $f(b) = a_0$ if $b \in B - A$, and $f(a) = a$ if $a \in A$. Clearly f is on B onto A, so $\kappa \leq {}^*\lambda$.

§2. Problem 1. 2.2(a) is obvious. In 2.2(b), let κ be a natural number and let X be any set such that $0 \in X$ and for every λ, if $\lambda \in X$ then $\lambda + 1 \in X$. By 2.1, $\kappa \in X$. Hence $\kappa + 1 \in X$ for any such X. So by 2.1, $\kappa + 1$ is a natural number.

Problem 2. One easily shows by induction (2.2(c)) that (*) for every n, $\mathcal{N}_n$— by (a), (b). (Note: 2.2(c) has a class variable $\mathcal{P}$ in it, and in proving * we take $\mathcal{N}$ for $\mathcal{P}$.) We now show that for any κ, if $\mathcal{N}_\kappa$ then κ is a natural number, by the 'induction' given in hypothesis (c)—taking $\mathcal{P}_\kappa$ to be 'If $\mathcal{N}_\kappa$ then κ is a natural number'. $\mathcal{P}_0$ is trivial since 0 is a natural number. Assume $\mathcal{N}_\kappa$ and assume: if $\mathcal{N}_\kappa$ then κ is a natural number. So κ is a natural number. Hence $\kappa + 1$ is a natural number (by 2.2(b)), so $\mathcal{P}(\kappa + 1)$ holds. So (*) is proved (by (c)), as desired.

Problems 3-12. The Corollaries are easy but require small arguments. For example: Prove 2.3(d)(i): Let $m \leq n$. By 1.4 there is a κ such that $m + \kappa = n$. By 1.3(a) $\kappa + m = n$, so $k \leq n$. Hence, by (b), κ is a natural number, proving 'at least one'. Suppose $m + k = n$ and $m + l = n$. Then $k + m = l + m$, so $k = l$ by (c). Thus 'at most one' is proved.

Also, we prove 2.3(d)(ii): Suppose $A \sim B$ and $B \subsetneq A$, where A is finite. Then $B \cup (A - B) = A$ and $B \cup \emptyset = B$, so $B \cup (A - B) \sim B \cup \emptyset$. Since B is finite, we can 'cancel' by (c), getting $A - B = \emptyset$, which is absurd.

The induction problems in 3-12, 13, and 14 more or less 'correct themselves'. Here is one: Problem 10 (prove 2.3(b)): Fix (A, R). We show by induction on 'n' that [for any $B \subseteq A$ of power n, if $B \neq \emptyset$ then B has a first element]. (We have just done more than half the work.) Write $\mathcal{P}(n)$ for $[\ldots n \ldots]$ above. $\mathcal{P}(0)$ holds vacuously, as $\emptyset \neq \emptyset$. Assume $\mathcal{P}(n)$. Suppose $B \subseteq A$, and $\overline{\overline{B}} = n + 1$, and $B \neq \emptyset$. Since $B \neq \emptyset$ we can choose $b \in B$. Then $B = \{b\} \cup (B - \{b\})$, a 'disjoint union'. By (b) $\overline{\overline{B - \{b\}}}$ is a natural number, say, k. Thus $k + 1 = n + 1$ so $k = n$ (by (c)). If $B - \{b\} = \emptyset$ then $B = \{b\}$, which has least element b. Suppose $B - \{b\} \neq \emptyset$. Then by the induction hypothesis (i.e., $\mathcal{P}(n)$), $B - \{b\}$ has a least element, c. It is easy to see that $\min\{b, c\}$ is a least element of B.

Problem 15. No induction is needed. $\Rightarrow$ is very easy. $\Leftarrow$ is a little bit harder. In both directions, use a result or two from 2.3(a)-(j).

§3. Problem 1. We want to prove that for *any* μ, ν, $(2 + \mu) + (2 + \nu) \leq (2 + \mu)(2 + \nu)$, i.e., $4 + \mu + \nu \leq 4 + 2\mu + 2\nu + \mu\nu$—which clearly holds.

Problem 2. *Hint.* First study the proof of 3.4(a) given in the book.

Problem 3. We must show that the F defined in (2) is into and onto $A^{B \times C}$ and one-to-one. We illustrate by showing onto (given into). Let $g \in A^{B \times C}$, that is, $g : B \times C \to A$. For each $c \in C$, let h_c be the function on B such that for all $b \in B$, $h_c(b) = g((b, c))$. Now let f be the function on C such that for any $c \in C$, $f(c) = h_c$. Then for any $b \in B$, and $c \in C$, $F(f)(b, c) = f(c)(b) = g((b, c))$; so $F(f) = g$.

Problem 4. Suppose $A)(B$. We will define F so that $A^{B \cup C} \underset{F}{\sim} A^B \times A^C$. Indeed, let F be the function on $A^{B \cup C}$ such that, for any $f \in A^{B \cup C}$, $F(f) = (f \restriction B, f \restriction C)$.

Problem 7. *Hint.* Take $\overline{\overline{A}} = x$ and consider $2 \times A$ and 2^A, if $\kappa \geq 3$ (the cases $\kappa = 0, 1, 2$ are trivial separately.

§4. Problem 1. $f(0) = g(0) = a$. Suppose $f(i) = g(i)$ and $i + 1 \leq q$. Then $f(i + 1) = \mathcal{A}_{f(i)} = \mathcal{A}_{g(i)} = g(i + 1)$. By induction, $f(i) = g(i)$ for all $i \leq q$.

Problem 2. Fix m. Take $\mathcal{A}_k = k + m$ for all k; and $a = 0$. Put $f_q =$ the unique f which works for q. f_q exists by 4.1(a) (which is applied several times below). Then (5) becomes: $m \cdot n = f_n(n)$. Then $m \cdot 0 = f_0(0) = 0$. Finally, $m \cdot (n+1) = f_{n+1}(n+1) = f_{n+1}(n) + m = f_n(n) + m$ (by Lemma (i) $= m \cdot n + m$.

§5. Problem 2. *Hint.* Take $\mathcal{P}(n)$ to be: if $m \neq n$ then $f(m) \neq f(n)$. (Here m is fixed.) Case $n = 0$ is easy. Suppose $\mathcal{P}(n)$ and suppose $m \neq n+1$. Easys if $m = 0$, so take $m = p+1$. Then $p \neq n$. But we cannot apply the induction hypothesis which involves m not p. Apparently we chose a $\mathcal{P}$ which does not work. Start over with a different $\mathcal{P}(n)$.

Problem 3. ((c) $\to$ (a)). By recursion, put $h(0) = a$ and $h(n+1) = f(h(n))$ for all n. We show by induction that for all n, $\mathcal{P}(n)$ holds, where

$$\mathcal{P}(n) \text{ is: for all } k \neq n,\ h(n) \neq h(k).$$

$\mathcal{P}(0)$ holds, as $a \notin \operatorname{Rng} f$ while $k \in \operatorname{Rng} f$ if $k \neq 0$. Assume $\mathcal{P}(n)$. Let $h(n+1) = h(k)$ where $k \neq n+1$. Then $k \neq 0$ as above. Say $k = l+1$. Thus $f(h(n)) = f(h(l))$. Since f is $1-1$, $h(n) = h(l)$, where $l \neq n$ (as $l+1 \neq n+1$). This contradicts the induction hypothesis. So $\mathcal{P}(n+1)$ holds. Now put $C = \operatorname{Rng} h \cup \{a\}$. We showed h is $1-1$ on N onto C, so $\overline{\overline{C}} = \aleph_0$. Thus $\overline{\overline{C}} \leq \overline{\overline{A}} = \kappa$, so $\aleph_0 \leq \kappa$, as desired.

CHAPTER 3

Problem 1. See any modern algebra book on the 'field of quotients' for the proof (of a more general result).

Problem 2. Similar to Problem 1.

Problem 3. *Hint.* Let $\underline{R}$ and $\underline{R}'$ be complete ordered fields. Apply the ideas in the hint in the book, Problem 3, to both $\underline{R}$ and $\underline{R}'$, getting W and W', and, by 2.2, a $(\underline{W}, \underline{W}')$-isomorphism h. Let $x \in R$. Put $D = \{w : w \in W$ and $w < x\}$. Clearly $D' = h[D]$ is non-empty and bounded above; so we can put $F(x) = \sup_{y \in D'} y$ (taken in $\underline{R}'$). It remains to show that F is the desired $(\underline{R}, \underline{R}')$-isomorphism. Be patient!

CHAPTER 4

§1. Problem 1. Recall that the phrase 'it is easy to see that' *always* means there are some details missing.

§2. Problem 1. By hypothesis, for each $i \in I$, $\{f : F(i) \underset{f}{\sim} G(i)\} \neq \emptyset$. By AC, there is a function h on I such that for each $i \in I$, $F(i) \underset{h(i)}{\sim} G(i)$. It is easy to check that $\bigcup_{i\in I} h(i)$ is a $1-1$ function on $\bigcup_{i\in I} F(i)$ onto $\bigcup_{i\in I} G(i)$, as desired.

Problem 3. As in Problem 1 above, let $F(i) \underset{h_i}{\sim} G(i)$ for all $i \in I$. let $\mathcal{T}$ be the function on $\prod_{i\in I} F(i)$ such that for all $f \in \prod_{i\in I} F(i)$, $\mathcal{T}(f)$ is the function on I such that for each $i \in I$, $\mathcal{T}(f)(i) = h_i(f(i))$. It is routine to show $\mathcal{T}$ is $1-1$ on $\prod_{i\in I} F(i)$ onto $\prod_{i\in I} G(i)$, as desired.

Problem 4. See Problems 1 and 7 of §3, Chapter 2 for special cases.

§3. Problem 1. *Hint.* Let $\overline{\overline{K_i}} = \kappa_i$ and $\overline{\overline{L_i}} = \lambda_i$ for each $i \in I$. Let $F : \bigcup_{i\in I} K_i \to \prod_{i\in I} L_i$. We show F is not onto. For each $i \in I$, let G_i be the function on K_i such that, for each $k \in K_i$, $G_i(k) = F(k)(i)$. Thus $G_i : K_i \to L_i$. But G_i is not onto L_i, since $\kappa_i < \lambda_i$ (and by 2.1.8(b)). Hence, by AC, there is an f on I such that for each $i \in I$, $f(i) \in L_i - G_i[K_i]$. Complete the proof by showing that $f \in \prod_{i\in I} L_i - F[\bigcup_{i\in I} K_i]$. (Hint^2. f has been chosen so that for each i, it differs at the ith place not from *one* function but from each of a whole group of functions.)

§4. Problem 4. $\mathfrak{c} \leq \overline{\overline{A}}$, as one see by considering the set of all constant functions on R to R. If $f, g \in A$ and $f(t) = g(t)$ for all $t \in Q$, then $f = g$. Indeed, if $x \in R$ then x can be expressed as $\lim_{n\to\infty} t_n$ where each $t_n \in Q$; since f and g are continuous, $f(x) = \lim_{n\to\infty} f(t_n) = \lim_{n\to\infty} g(t_n) = g(x)$. Since each $f \in A$ is thus uniquely determined by $f \restriction Q \in R^Q$ it is easy to show $\overline{\overline{A}} \leq \overline{\overline{R^Q}} = \mathfrak{c}^{\aleph_0} = \mathfrak{c}$.

Problem 5. Problem 5 is harder than 4 or 6. Try it.

Problem 6. *Hint.* $U \leq R$ is open iff it is a countable union of open intervals of the form (s, t) wherer $s, t \in Q$ and $s < t$.

Problem 7. Denote $\mathcal{P}(N)$ by $\mathcal{S}$. Since $\mathcal{S} \sim R$ we can assume 'R' in Problem 7 has been replaced by '$\mathcal{S}$'. Suppose $A \subseteq \mathcal{S}$ and $N \underset{F_0}{\sim} A$ and put $\kappa = \overline{\overline{\mathcal{S} - A}}$. We define $W_0 = \{n \in N : n \notin F_0(n)\}$. As in Cantor's proof (cf. 3.1), $W_0 \in \mathcal{S} - A$. Define F_1 on N by: $F_1(0) = W_0$, $F_1(n+1) = F_0(n)$. Clearly $N \underset{F_1}{\sim} A \cup \{W_0\}$. Using recursion we can iterate this whole process to define W_i for each $i \in N$, where the W_i's are distinct members of $\mathcal{S} - A$. Hence $\aleph_0 \leq \kappa$. Say $\kappa = \aleph_0 + \lambda$. It follows that $\mathfrak{c} = \aleph_0 + \kappa = \aleph_0 + \aleph_0 + \lambda = \aleph_0 + \lambda = \kappa$. So $\overline{\overline{\mathcal{S} - A}} = \mathfrak{c}$, as was to be proved.

CHAPTER 5

§1. Both problems are pretty easy, but they are important and should be done.

§2. Problem 1. Prove 2.3(a). Let $(\sigma_i : i \in I)$ be a function on I whose values σ_i are order types. We can form $(\overline{\overline{\sigma_i}} : i \in I)$. By Assumption 4.2.1 there is a (fixed) list $(A_i : i \in I)$ such that for each $i \in I$, $\overline{\overline{A_i}} = \overline{\overline{\sigma_i}}$. For any $i \in I$, put $C_i = \{(A_i, <) : i \in I$ and $(A_i, <)$ is an order of type $\sigma_i\}$. For each $i \in I$, $C_i \neq \emptyset$ (clearly) and C_i is 'set' (easily). So, by AC, there is a function $\underline{B}$ on I such that for each $i \in I$, $\underline{B}_i \in C_i$, so $T_p\underline{B}_i = \sigma_i$. This proves 2.3.

Problem 3. Clearly $(N - \{0\}, \mathfrak{c})$ and $(N, \mathfrak{c})$ (where $\mathfrak{c}'$ is $\mathfrak{c}$ 'cut down') have type ω, and $1 + \omega$, respectively; and are isomorphic. So $1 + \omega = \omega$.

Problem 4. Clearly $1+\omega \neq \omega+1$. Clearly $2\cdot\omega = \omega$ while $\omega\cdot 2 = \omega+\omega \neq \omega$.

Problem 6. Take $\alpha = \beta = 1$, $\gamma = \omega$. Then $(\alpha + \beta)\gamma = 2 - \omega = \omega$, while $\alpha\gamma + \beta\gamma = \omega + \omega \neq \omega$.

Problems 5 and 7. Problems 5 and 7 are easy but good.

Problem 8. Starting hint. Using 2.5(a), one obtains an isomorphism between B (and its induced order) and $(Q, \leq)$.

CHAPTER 6

§1 has no problems.

There are good hints in the book for the problem in §2.

§3. **Problem 1.** $(\mathcal{A}_x : x \in X) = \{(x, \mathcal{A}_x) : x \in X\}$, which exists as a case of $\{B_x : x \in U\}$ whose existence has been proved in item 2 in our list.

Problems 2 and 3 are the key ones.

Problem 2. There are two quite different proofs. The first depends on the Replacement Axiom but not on the Power Set Axiom: Let $b \in B$. Put $A \times \{b\} = \{(a, b) : a \in A\}$ which exists by items 2 and 8 in our list. Now, $A \times B = \bigcup\{A \times \{b\} : b \in B\}$, which exists by items 2 and 3 in our list. The second proof depends on the Power Set Axiom but not on the Replacement Axiom: Note that $\{a\}, \{a, b\} \in \mathcal{P}(A \cup B)$—if $a \in A$ and $b \in B$. So if $a \in A$ and $b \in B$ then $(a, b) = \{\{a\}, \{a, b\}\} \in \mathcal{PP}(A \cup B)$. So $A \times B = \{z : z \in \mathcal{PP}(A \cup B)$ and for some $a \in A$ and $b \in B, z = (a, b)\}$ which exists by the Separation Axiom.

Problem 3. Hints. Again there are two proofs. The first depends on Replacement but not on the Union Axiom. The second depends on Union but not on Replacement.

Problems 4 and 5 are routine, or soon will be.

CHAPTER 7

§1. Problem 3. Let $f, g \in H$. Put $U = \operatorname{dom} f$ and $V = \operatorname{dom} g$. Both U and V are initial segments of $\underline{A}$. Since $\underline{A}$ is connected, one can easily show (by considering $U \cap V$) that one of U, V, say U, is an initial segment of the other, V. Since $g \in H$, g is an isomorphism of $\underline{V}$ onto an initial segment of $\underline{B}$. Since 'isomorphism preserves everything', $g[U]$ is an initial segment of $g[V]$, so of $\underline{B}$. Thus both f and $g \restriction U$ are isomorphisms of $\underline{U}$ onto initial segments of $\underline{B}$. So $f = g \restriction U$, by 1.1(d), and hence $f \subseteq g$.

Problem 4. Partial solution. First we show: h is a function. Clearly h is a set of ordered couples. Suppose $(a, b), (a, b') \in h$. Then $(a, b) \in f$ and $(a, b') \in f'$, for some $f, f' \in H$. Since H is a chain, either $f \subseteq g$ or $g \subseteq f$, say the former. Then $(a, b), (a, b') \in g$, so $b = b'$.

Using several similar arguments one may show also that: $\operatorname{Dom} h$ is an initial segment of A, $\operatorname{Rng} h$ is an initial segment of $\underline{B}$, h is $1 - 1$, and for any $a_1, a_2 \in \operatorname{Dom} h$, $a_1 \underset{\underline{A}}{<} a_2$ iff $h(a_1) \underset{\underline{B}}{<} h(a_2)$.

Problem 7. Fix $b \in B$. Assume that for each $c < b$, there is exactly one f which works for $\underline{B}_c$ (call it f_c). We will show there is an f which works for $\underline{B}_b$. (By (6) there is always at most one such f.) By (5),

$$\text{(a) if } c < d < b \text{ then } f_c = f_d \restriction \operatorname{pred} c.$$

Consider first the case in which b is a limit element or the first element. Let f be on $\operatorname{pred} b$ with $f(c) = f_{c^+}(c)$ for all $c < b$. (c^+ is the immediate successor of c.) Then $f(c) = f_{c^+}(c) = \mathcal{A}_{f_{c^+}\centerdot\operatorname{pred} c}$ (as f_{c^+} works for $\underline{B}_c$) $= \mathcal{A}_{f_c}$ (as, by (5), $f_c = f_{c^+} \restriction \operatorname{pred} c$). Thus f works for $\underline{B}_b$, as desired. The other case is where $b = c^+$. This time put $f = f_c \cup \{(c, \mathcal{A}_{f_c})\}$. Clearly f is a function on $\operatorname{pred} b$, and also (b) $f_c = f \restriction \operatorname{pred} c$. If $d < c$, then $f(d) = f_c(d) = \mathcal{A}_{f_c\centerdot\operatorname{pred} d} = \mathcal{A}_{f\centerdot\operatorname{pred} d}$ as desired. Finally, $f(c) = \mathcal{A}_{f_c} = \mathcal{A}_{f\centerdot\operatorname{pred} c}$, by (b). Thus f works for $\underline{B}_b$.

Problem 8. Let $\underline{A}$ be any well-order. By 1.5, there is exactly one f which works for $\underline{A}$ over $\mathcal{B}$. Clearly, by our definition of $\mathcal{O}$, $\mathcal{O}(\underline{A}) = \mathcal{B}_f$. We want to prove that (7) holds for our $\mathcal{O}$ and $\underline{A}$, i.e. $\mathcal{O}(\underline{A}) = \mathcal{B}_{\langle \mathcal{O}(\underline{A}_a) : a \in A \rangle}$. Thus it is enough to show $f(a) = \mathcal{O}(\underline{A}_a)$, for each $a \in A$. Since f works for $\underline{A}$ over $\mathcal{B}$, we have: $f(a) = \mathcal{B}_{f \restriction \mathrm{pred}\, a}$. By definition, $\mathcal{O}(\underline{A}_a) = \mathcal{B}_g$ where g works for $\underline{A}_a$ over $\mathcal{B}$. Thus it is enough to prove that $f \restriction \mathrm{pred}\, a = g$. But this holds by (5) and (6) (and (0)).

Problem 9. Using (0) for $\underline{A}$ and for $\underline{A}'$, we see it is enough to show:

$$\text{If } \underline{A} \underset{f}{\cong} \underline{A}' \text{ then } (*) \text{ for any } a \in A,\ \mathcal{O}(\underline{A}_a) = \mathcal{O}(\underline{A}'_{f(a)}).$$

(To this it is enough to apply (0) to (the original) $\underline{A}$ and $\underline{A}'$.) Assume $\underline{A} \underset{f}{\cong} \underline{A}'$. We prove (*) by $\underline{A}$-induction on 'a'. Suppose $\mathcal{O}(\underline{A}_b) = \mathcal{O}(A'_{f(a)})$ for all $b < a$. Then

$$\begin{aligned} \mathcal{O}(\underline{A}_a) &= \mathcal{A}_{\{\mathcal{O}(\underline{A}_b):b<a\}} \text{ (by (7), since if } b < a \text{ then } (\underline{A}_a)_b = \underline{A}_b) \\ &= \mathcal{A}_{\{\mathcal{O}(\underline{A}'_{f(a)}):b<a\}} \text{ (by the inductive hypothesis)} \\ &= \mathcal{A}_{\{\mathcal{O}(\underline{A}'_x):x<f(a)\}} \text{ (as pred } f(a) = f[\mathrm{pred}\, a]) \\ &= \mathcal{O}(\underline{A}'_{f(a)}) \text{ (by (7) with } \underline{A}'_{f(a)} \text{ for "}\underline{A}\text{"), as was to be proved.} \end{aligned}$$

§2. Problem 1. Let K be any set of ordinals. We will show there is a β such that for any $\alpha \in K$, $\alpha \le \beta$. Put $\underline{B}$ = the ordered sum $\sum_{\alpha \in (K,a)} \underline{W}(\alpha)$. (Note that $\underline{W}_\alpha = \underline{W}(\alpha)$.) $\underline{B}$ is a well-order by 5.1.1(a); put $\beta = \mathrm{Ord}\, \underline{B}$. By 1.3, for any $\alpha \in K$, $\underline{W}(\alpha)$ is isomorphic to an initial segment of $\underline{B}$, so $\alpha = \mathrm{Ord}\, \underline{W}(\alpha) \le \mathrm{Ord}\, \underline{B} = \beta$, as desired.

Problem 2. (Of course, $\varepsilon_x = \{(\mu, \nu) : \mu, \nu \in x \text{ and } \mu \in \nu\}$; and well-order, here means strict well-order.) $\Rightarrow$ follows at once from results in 2.12-13 and 2.7-8.

Proof of $\Leftarrow$: For readability write $A = x$. Assume A is a transitive set and $\underline{A} = (A, \varepsilon_A)$ is a (strict) well-order. Put $\alpha = \mathrm{Ord}\, \underline{A}$ (cf. 2.2). By 2.8(b) plus 2.12-13, $\underline{A} \underset{f}{\cong} (\alpha, \varepsilon_\alpha)$ where $f(a) = \mathrm{Ord}\, \underline{A}_a$ for each $a \in A$. We will prove by $\underline{A}$-induction on 'a' that $f(a) = a$ for all $a \in A$. Suppose $a \in A$ and $f(b) = b$ for all $b \in_A a$. Then

$$\begin{aligned} f(a) &= \mathrm{Ord}\, \underline{A}_a = \{\mathrm{Ord}\, \underline{A}_b : b \in \mathrm{pred}_{\underline{A}}\, a\} \text{ (by 2.2, as } (\underline{A}_a)_b = \underline{A}_b \text{ when } b <_{\underline{A}} a) \\ &= \{\mathrm{Ord}\, \underline{A}_b : b \in_A a\} \text{ (at once)} \\ &= \{b : b \in_A a\} \text{ by the induction hypothesis} \\ &= a \text{ (since } a \text{ is transitive).} \end{aligned}$$

Thus $f = \mathrm{Id}_A$ and then $A = \alpha$, so A is an ordinal, as was to be proved.

§3. Problem 1. We call I an ideal in A if $1 \notin I, 0 \in I, x - y \in I$ whenever $x, y \in I$, and $ax \in I$ whenever $x \in I$ and $a \in A$. Assume I_0 is an ideal (in A). We want to show A_0 is included in a maximal ideal. Let K be the family of all ideals $I \supseteq I_0$. Let C be any chain of members of I. Put $I^* = \bigcup_{I \in C} I$. It is easy to check that I^* is an ideal and $I_0 \subseteq I^*$. Thus $I^* \in K$. By Zorn's Lemma (with $(K, \subseteq_K)$ for the $\underline{A}$ *there*), K has a maximal member I. If I' is any ideal and $I \subseteq I'$, then obviously $I = I'$. So I is as desired.

Problem 2. Hints given.

Problem 3. *Hint.* Show $\leq$ can be ($\subseteq$) extended to a maximal partial order $\leq'$ for A. Then show $\leq'$ must be an order.

§5. Problem 2. There are several proofs of 5.4. Here are hints for one. First show that for any α, $\omega_\alpha \geq \alpha$. (One can imitate the proof of 1.1(a), but over the class of all ordinals.) Thus $\aleph_\kappa \geq \kappa$, so we can let α be the least ordinal such that $\aleph_\alpha \geq \kappa$. Suppose $\aleph_\alpha > \kappa$. Each of the two cases (α limit or not) easily leads to a contradiction.

Problem 4. *Hint for the Hint*: By recursion over ω, put

$$\begin{cases} \kappa_0 = \omega \\ \kappa_{n+1} = \omega_{\kappa_n} \text{ for all } n. \end{cases}$$

Take $\sup\{\kappa_n : n \in \omega\}$. Continue.

§6. Problem 2. First prove easily the general facts:

(A) If Q is a set of ordinals then $\sup_{\alpha \in Q} \alpha = \bigcup_{\alpha \in Q} \alpha$.

(B) $\overline{\overline{\bigcup_{i \in I} A_i}} \leq \sum_{i \in I} \overline{\overline{A_i}}$ (which was needed for the second sentence in 6.4). Hint: If $i \in I$ and $a \in A_i$, put $F((i, a)) = a$. F maps $\bigcup_{i \in I} (\{i\} \times A_i)$ onto $\bigcup_{i \in I} A_i$. Apply 2.1.8(b). (Compare 4.4.4 and 4.4.2.)

(C) If κ is infinite, then κ is a limit ordinal. (By, for example, 6.2(b), $\kappa = \kappa + 1$. Now one easily shows that for any α, $x \neq \alpha \cup \{\alpha\}$.)

Proof of 6.5(a): Put $\lambda = \sum_{\alpha \in Q} \overline{\overline{\alpha}} \leq \sum_{\alpha \in Q} \kappa \leq \kappa \cdot \kappa = \kappa$. It remains to prove $\kappa \leq \lambda$. First, suppose κ is a limit cardinal (i.e., κ is a limit element in the class of cardinals, or, equivalently, for some limit ordinal δ, $\kappa = \aleph_\delta$). For each (cardinal) $\nu < \kappa$, there is an $\alpha \in Q$ such that $\nu \leq \alpha$ and so $\nu \leq \overline{\overline{\alpha}} \leq \lambda$. Since κ is a limit cardinal, $\kappa = \bigcup_{\nu < \kappa} \nu \leq \lambda$. Finally, suppose $\kappa = \mu^+$. We claim $\overline{\overline{Q}} = \kappa$. If not, $\overline{\overline{Q}} < \kappa$, so $\overline{\overline{Q}} \leq \mu$. Hence $\kappa = \bigcup_{\alpha \in Q} \alpha$ (by (A)) $= \overline{\overline{\bigcup_{\alpha \in Q} \alpha}} \leq \sum_{\alpha \in Q} \overline{\overline{\alpha}}$ (by (B)) $\leq \overline{\overline{Q}} \cdot \mu \leq \mu \cdot \mu = \mu$, which is absurd. Thus $\overline{\overline{Q}} = \kappa$, as claimed. Hence $\lambda = \sum_{\alpha \in Q} \overline{\overline{\alpha}} = \sum_{\alpha \in Q'} \overline{\overline{\alpha}}$ (where $Q' = Q - \{0\}$) $\geq \sum_{\alpha \in Q'} = \overline{\overline{Q'}} = \overline{\overline{Q}}$ (as Q is infinite by (C)) $= \kappa$. Thus (a) is proved.

Proof of 6.5(b): If $\mu \in Q$, put $I_\mu = \{i \in I : \overline{\overline{I_i}} = \mu\}$. Suppose $\kappa \neq \sup_{i \in I} \lambda_i =$ (say) λ. Then $\kappa = \sum_{i \in I} \lambda_i = \sum_{\mu \in Q} \sum_{i \in I_\mu} \mu = \sum_{\mu \in Q} \mu \cdot \overline{\overline{I_\mu}} \leq \sum_{\mu \in Q} \lambda \cdot \overline{\overline{I}} = \overline{\overline{Q}} \cdot \lambda \cdot \overline{\overline{I}} < \kappa$ (as κ is infinite and $\overline{\overline{Q}} \leq \overline{\overline{I}}$). This is absurd, so (b) holds.

CHAPTER 8

§1. Problem 1. 1.1(a),(b),(c) are excellent exercises in making proofs by transfinite induction (each is easy, but not too easy).

Problem 2. Prove 1.2(a),(b),(c). (a) and (b) can easily be done (without induction). Proof of (c):

By the class version of 7.1.1(a) (see the solution of Problem 2 of §7.5) we obtain $\mathrm{Rk}\,\alpha \geq \alpha$. We prove $\mathrm{Rk}\,\alpha \leq \alpha$ by induction on α. The case $\alpha = 0$ is trivial. Suppose $\alpha = \delta$, a limit ordinal, and $\gamma \subseteq V_\gamma$ for all $\gamma < \alpha$. Then $\delta = \bigcup_{\alpha<\delta} \alpha \subseteq \bigcup_{\alpha<\delta} V_\alpha = V_\delta$. Finally, suppose $\alpha = \beta + 1$ and $\beta \subseteq V_\beta$. Then $\beta \in P\beta \subseteq PV_\beta = V_{\beta+1} = V_\alpha$. Since V_α is transitive, $\beta \subseteq V_\alpha$, so $\alpha = \beta \cup \{\beta\} \subseteq V_\alpha$.

Hence, in general $\alpha \subseteq V_\alpha$, so $\alpha \in V_{\alpha+1}$. Hence V_α holds and $\mathrm{Rk}\,\alpha \leq \alpha$. We already saw that $\mathrm{Rk}\,\alpha \geq \alpha$, so $\mathrm{Rk}\,\alpha = \alpha$, and 1.2(c) is proved.

Problem 3. Fix x. Define W_n by:

$$W_0 = x$$
$$W_{n+1} = \bigcup(W_n) \ (= \textstyle\bigcup_{t\in W_n} t).$$

Since ω exists, $\{W_n : n \in \omega\}$ exists by Replacement and hence $y = \bigcup_{n\in\omega} W_n$ exists (by Union). Clearly $x \subseteq y$. Let $u \in v \in y$. Then $v \in W_n$ for some n. Hence $u \in \bigcup(W_n) = W_{n+1} \subseteq y$. So y is transitive (in fact, one easily shows y *is* $Tc(x)$).

Problem 4. 1.7(a) is easily proved. Proof of 1.7(b). Suppose (2) facts. Thus there exists $A \neq \emptyset$ such that A has no ε-minimal element. For each $x \in A$, $(\varepsilon\text{-Pred } x) \cap x \neq \emptyset$. By Replacement $\{(\varepsilon\text{-Pred } x) \cap A : X \in A\}$ exists, so by Choice it has a choice function F. Pick $x \in A$. By ordinary recursion (and Replacement) define f by

$$\begin{cases} f(0) = x \\ f(n+1) = F(\mathrm{Pred}\, f(n) \cap A). \end{cases}$$

Clearly f is an ε-descending sequence, so (3) fails.

§2. Problem 1. One can easily check that

(A) Rk x is finite if for some n, $x \in V_n$.

(B) $Tc\{x\} = \{x\} \cup Tcx$

Still fairly simple is the important

(C) Rk $x = \sup_{y \in x}(\text{Rk}\, y + 1)$

In (C), $\leq$ is nothing. For $\geq$, you might consider two cases: $\{\text{Rk}\, y + 1 : y \in x\}$ has a largest member, and 'otherwise'.

(2.4) ($\Leftarrow$): By a simple induction, one shows each V_n is finite. Now suppose $x \in V_n$. Then $\{x\} \subseteq V_n$ so $Tc\{x\} \subseteq V_n$. Thus if $y \in Tc\{x\}$ then $y \in V_n$, so $y \subseteq V_n$ and y is finite.

($\Rightarrow$): Let (α) be the statement: For every x of rank $< \alpha$, if every member of $Tc\{x\}$ is finite then Rk x is finite. We will prove by induction on 'α' that for every α, (α) holds. Assume (α) holds for all $\alpha < \beta$. Suppose Rk $x = \alpha < \beta$ and (*) each member of $Tc\{x\}$ is finite. Suppose $y \in x$. Then Rk $y <$ Rk $x < \beta$, so Rk $y < \alpha$. Also, every member z of $Tc\{y\}$ ($= \{y\} \cup Tcy$ (by B)) is finite. Indeed, if $z = y$ then $y \in x \subseteq Tcx \subseteq Tc\{x\}$, so $z = y$ is finite by (*). Otherwise, $z \in Tcy \subseteq Tcx \subseteq Tc\{x\}$, so again z is finite, by (*). It now follows from the induction hypothesis that Rk y is finite. Thus Rk $x = \sup_{y \in x}(\text{Rk}\, y + 1)$ (by (C)), where x is a finite set, and each Rk $y + 1$ (for $y \in x$) is finite. Hence Rk x is finite. By induction, (α) holds for all α, i.e., ($\Rightarrow$) holds, as desired.

§3. Problem 1. By 7.1.2 one of t and t' is isomorphic to an initial segment of the other. Say f is an isomorphism of t onto t'', an initial segment of t'. We claim that $f(U) = U$ for all $U \in t$. (Clearly the theorem follows.) The claim will be proved by t-induction on 'U'. Suppose $f(U) = U$ for all $U \subsetneq V$, where $V \in t$. If V is first in t, then by (iii) $V \neq \emptyset$, and likewise $f(V) = \emptyset = V$. Suppose there in a largest $U \in t$ such that $U \subsetneq V$. Then by (ii), $V = P(U)$. Also $f(V)$ is the immediate successor of $f(U)$ in t'' and hence, clearly, in t'. Hence by (ii), $f(V) = P(f(U)) = P(U) = V$, as desired. Now let V be a limit element in t. Note that (*) $W \in t$ and $W \subsetneq V$ iff $f(W) \in t'$ and $f(W) \subsetneq f(V)$ iff $W \in t'$ and $W \subsetneq f(V)$. Since V is a limit element, $f(V)$ is a limit element in t' (or t'') so, by (iii),

$$f(V) = \bigcup_{\substack{W \in t' \\ W \subsetneq f(V)}} W = \bigcup_{\substack{W \in t \\ W \subsetneq V}} W \text{ (by (*))} = V \text{ (by (iii))},$$

completing the proof.

CHAPTER 9

§1. Problem 1. We prove by induction on Φ, that:

No proper initial segment of a formula Φ is a formula.

If Φ is an atomic formula, say Qxy (Q 2-place), then clearly no proper initial segment of Φ is a formula. Let us write $\mathcal{P}(\Phi)$ for 'no proper initial segment of Φ is a formula'. Suppose $\mathcal{P}(\Theta)$ and $\Phi =\sim \Theta$. Clearly $\sim$ is not a formula, so any proper initial segment of Φ which is a formula is of the form $\sim X$. But clearly any formula (like $\sim X$) which starts with $\sim$ is of the form $\sim \Theta'$, where $X = \Theta'$ is a proper initial segment of Θ, contrary to hypothesis. The case where $\Phi = \forall x \Theta$ is similar. Now suppose $\mathcal{P}(\Theta)$ and $\mathcal{P}(\Theta')$ and $\Phi = (\Theta \to \Theta')$ and assume Φ' is a proper initial segment of Φ. Clearly (using induction) any formula, like Φ', starting with (is of the form $(\Psi \to \Psi')$. If $l(\Psi) < l(\Theta)$ then Ψ is a proper initial segment of Θ, absurd. If $l(\Psi) = l(\Theta)$ then clearly Ψ' is a proper initial segment of Θ', absurd. So we must have $l(\Psi) > l(\Theta)$. Thus the ('main') occurrence of $\to$ in $(\Theta \to \Theta')$ must be in Ψ. But again, clearly (by induction) such an occurrence of $\to$ in a formula Ψ must have a formula, α, just to its left and another formula, β, just to its right (all inside Ψ). But then β is a proper initial segment of Θ', a contradiction, and the proof is complete.

Problem 2. First read the nine lines of hints starting at "Proof". Note: The proof that logical axioms satisfy $(*)$ divides into inner axioms and showing that if Φ satisfies $(*)$ so does $\forall x \Phi$.

Problem 3. The last sentence before 4.4 should end with: "is deductively closed *provided* $Ax \vdash \exists x \Psi$; that is:". Also, 4.4(a) should begin: "Suppose $Ax \vdash \exists x \Psi$. If..."

Assume all the hypotheses of (the new) 4.4(a). By 4.3 and our hypotheses $\vdash \exists x Px \to (\Sigma_1 \wedge \cdots \wedge \Sigma_n \to \Sigma)^{(P)}$. By considering the definition of $\Phi^{(P)}$, one easily sees $(\Sigma_1 \wedge \cdots \wedge \Sigma_n \to \Sigma)^{(P)}$ is identical with $\Sigma_1^{(P)} \wedge \cdots \wedge \Sigma_n^{(P)} \to \Sigma^{(P)}$, so we have $\vdash \exists x Px \to (\Sigma_1^{(P)} \wedge \cdots \wedge \Sigma_n^{(P)} \to \Sigma^{(P)})$. But then clearly $\vdash \exists x Px \wedge \forall x(Px \leftrightarrow \Psi) \to \Sigma_1^{*\Psi} \wedge \cdots \wedge \Sigma_n^{*\Psi}$. By hypothesis, $Ax \vdash \Sigma_1^{*\Psi} \wedge \cdots \wedge \Sigma_n^{*\Psi}$, hence $Ax \vdash \forall x(Px \leftrightarrow \Psi) \to \Sigma^{*\Psi}$. Hence clearly $Ax \vdash \Sigma^{*\Psi}$, as desired.

CHAPTER 10

§1. Problem 1. Preliminary remarks needed: We say Φ is $Z_0^{(0)}$-bounded if Φ is equivalent in $Z_0^{(0)}$ to a bounded formula. (Note: The first formula in (1)(a), $u \subseteq v$, is trivially $Z_0^{(0)}$-bounded. In fact, since $Z_0^{(0)} \vdash x \subseteq y \leftrightarrow (\forall t \in x)\, t \in y$, $x \subseteq y$ is $Z_0^{(0)}$-bounded in the stronger sense, where only $\in$-bounded quantifications occur. But we do not need to discuss this sense.) If Φ is $Z_0^{(0)}$-bounded, then so are $(\forall u \in v)\Phi$ and $(\forall u \subseteq v)\Phi$ and clearly also $(\exists u \in v)\Phi$ and $(\exists u \subseteq v)\Phi$. We often consider an iota-formula Φ (e.g. "$z = (x, y)$" or "$y = Px$"), which is not strictly speaking a formula of $Z_0^{(0)}$ since (way back in Chapter 1, for example) their definition used the ι-symbol. We understand that the ι-symbol can be removed as on line 2, page 100, to get a $Z_0^{(0)}$-formula Φ' and in say Φ is equivalent in $Z_0^{(0)}$ to bounded formula if and only if Φ' is. This usage is understood on lines 5-9 of §1. Another example is this: The formula "$B = \{x : x \in A \wedge \Phi\}$" is $Z_0^{(0)}$-bounded if Φ is. In fact, it (or its $Z_0^{(0)}$-translate) is $Z_0^{(0)}$-equivalent to $(\forall x \in B)(x \in A \wedge \Phi) \wedge (\forall x \in A)(\Phi \rightarrow x \in B)$.

Now we deal with $z = (x, y)$ in 1(a). First note that

$$Z_0^{(0)} \vdash v = \{x, y\} \leftrightarrow x \in v \wedge y \in v \wedge (\forall t \in v)(t = x \vee t = y)$$

of course $v = \{x\}$ is also $Z_0^{(0)}$-bounded. Thus

$$Z_0^{(0)} \vdash z = (x, y) \leftrightarrow (\exists u \in z)(\exists v \in z)(u = \{x\} \wedge v = \{x, y\}),$$

so $z = (x, y)$ is $Z_0^{(0)}$-bounded.

The reader should do 1(b). This will be rather long, but the main requirement is patience. (In 1(b) you will require for the first time the use of $\subseteq$-bounded quantification.)

Problem 2 will now be straightforward.

Problem 3. The last three lines of (2) should be headed by '(b)'. For 2(a) we first check that if u is a partial universe, so that for some tower t, $u \in t$, then for some tower t', $t' \in PPu$ and $u \in t'$. Indeed, take $t' = \{v \in t : v \subseteq u\}$. Clearly $t' \subseteq Pu$, so $t' \in PPu$.

Now we want to show that 'u is a partial universe' is $Z_0^{(0)}$-absolute. There are two cases in the definition of absolute. We do the one about w, and leave the one about V to the reader. We must show (working in $Z_0^{(0)}$) that: if (i) [w is a limit partial universe and $u \in w$], then (ii) {[there exists t such that t is a tower and $t \in PPu$ and $u \in t$] iff [there exists t such that $t \in w$ and (t is a tower)$^{(w)}$ and $(t \in PPu)^{(w)}$ and $(u \in t)^{(w)}$]}. The three formulas relativized to w are all bounded. Hence (by 1.1) the $^{(w)}$'s can be deleted. Clearly it is enough to show that if (i) holds, t is a tower, $t \in PPu$, and $u \in t$ then $t \in w$. But, by 8.3.2 and our hypotheses, it is clear that the partial universes u, PPu, and w must lie in that ($\subseteq$)-order, so $t \in w$ (as $t \in PPu$).

Problem 6. Additional hints. By definition, u is a partial universe iff for some t, t is a tower and $u \in t$. The trouble is thatthis 'for some t' is not a bounded quantification. Try instead (perhaps differently in several cases) things *like* $(\exists t' \in u)(t'$ is a tower $\wedge (\exists v \in t)(u = PPt'))$.

§2. Problem 1. Assume $ZFC^{(0)}$ is consistent. We will prove that $ZFC^{(0)}+$ 'there is no inaccessible cardinal $> \aleph_0$' is consistent. Write 'u is a big universe' or $\Delta(u)$ for 'u is a limit universe and not the first, and u has class replacement'. There exists an inaccessible cardinal $> \aleph_0'$ and $\exists u \Delta(u)$ are equivalent in $ZFC^{(0)}$ (by 8.2.6 and 8.3.3). It remains to show that $ZFC^{(0)} + \sim \exists u \Delta(u)$ is consistent. This is certainly so if $ZFC^{(0)} \vdash \sim \exists u \Delta(u)$. So we may assume that contrary, i.e., the theory $T = ZFC^{(0)} + \exists u \Delta(u)$ is consistent.

Let $\Psi(x)$ be '$x \in$ the first big universe'. Clearly $T \vdash \exists x \Psi(x)$. By 1.2–1.4 (always ignoring the case 'P is V'), we see that if Σ is any axiom of $ZFC^{(0)}$ then T (or less) $\vdash \Sigma^{(\Psi)}$. We finish by showing that $T \vdash (\sim \exists u \Delta(u))^{(\Psi)}$. (If so, then Ψ interprets $ZFC^{(0)} + \sim \exists u \Delta(u)$ in T, so $ZFC^{(0)} + \sim \exists u \Delta(u)$ is consistent, completing the proof of 2.2.) Indeed, it follows from (2) and (3) that $\Delta(u)$ is T-absolute. Hence $T \vdash \Psi(u) \wedge (\Delta(u))^{(\Psi)} \rightarrow \Delta(u)$. But $\Psi(u)$ is 'u is the first big universe', while $\Delta(u)$ is '$u \in$ the first big universe'. Thus $T \vdash \sim (\Psi(u) \wedge (\Delta(u))^{\Psi})$, so $T \vdash \sim \exists u(\Psi(u) \wedge (\Delta(u))^{(\Psi)})$, and $T \vdash (\sim \exists u \Delta(u))^{(\Psi)}$, as was to be shown.

CHAPTER 11

§1. Problem 1. *should* say: Prove 1.4(a),(b),(c). Hints for (a): Let $\lambda < \mathrm{cf}\,\kappa$. As in the book, if $f : \lambda < \kappa$ then $f : \lambda \to \alpha$, for some $\alpha < \kappa$. Hence $\kappa^\lambda \le \sum_{\alpha<\kappa} \overline{\overline{\alpha}}^\lambda$. Also note that by G.C.H., if $\mu, \mu' < \kappa$ and $\nu = \max(\mu, \mu')$ is finite, then $\mu^{\mu'} \le \nu^\nu = \nu^+ \le \kappa$. Now complete (a); and do (b) and (c), which are easy.

Problem 2. 1.5(a) should begin: Any *infinite* order ....

Proof of 1.5(a). First we prove: *any order $\underline{A}$ has a cofinal subset which is a well-order (under $<^{\underline{A}}$).* Pick $b \notin A$ and well-order A by $\prec$. Let $\overline{\overline{A}} = \kappa$. We define a_η for $\eta < \kappa^+$ by recursion: If $(a_\eta : \eta < \xi)$ is $<^{\underline{A}}$-increasing and (its range) is not cofinal in $\underline{A}$, then let a_ξ be the $\prec$-first member of A such that $a_\eta <^{\underline{A}} a_\xi$ for all $\eta < \xi$. Otherwise (for 'purely technical reasons'), put $a_\xi = b$. Clearly $a_\xi = b$ for some $\xi < \kappa^+$ (else $\kappa = \kappa^+$); let γ be the first such ξ. Thus $\{a_\eta : \eta < \gamma\}$ is a cofinal subset of $\underline{A}$ which is well-ordered by $<^{\underline{A}}$ (indeed, in order type γ).

Now we complete (a) by proving: *Any well-order $\underline{B}$, of power $\kappa \ge \aleph_0$, has a cofinal subset $\underline{C}$ of ordinal $\le \kappa$.* Clearly we can assume $\underline{B}$ is (ζ, ϵ_ζ) where $\kappa \le \zeta < \kappa^+$. Write $\zeta = \{\alpha_0, \ldots, \alpha_\xi, \ldots\}_{\xi<\kappa}$. We now define by recursion ordinals β_ξ for $\xi < \kappa$. If there is an ordinal γ in ζ which is greater than β_η for all $\eta < \xi$, then let γ' be the first such γ and put $\beta_\xi = \max(\gamma', \alpha_\xi)$; if not put $\beta_\xi =$ (say) ζ. If $\beta_\xi < \zeta$ for all $\xi < \kappa$, then $\{\beta_\xi : \xi < \kappa\}$ is cofinal in ζ (since $\beta_\xi \ge \alpha_\xi$) and has order type κ. Otherwise, let ξ be the first ordinal such that $\beta_\xi = \zeta$. Then $\{\beta_\eta : \eta < \xi\}$ is cofinal in ζ and has order type $\xi < \kappa$.

Problem 3. The proof of 1.5(b) consists of several arguments. 1.5(a) is applied twice near the beginning.

No induction or recursion is needed.

Problem 4. Prove 1.5(c). The author cannot find a 'proof' as simple as the book thinks it knows (though one may exist). Here is an outline (to be completed) of a (well-known) proof: First the reader should give the easy proof of

Proposition (I). *The following are equivalent:*

(i) *α is a regular cardinal*

(ii) $\alpha = \mathrm{cf}\,\alpha$

(iii) *for some* β, $\alpha = \operatorname{cf}\beta$.

Theorem (II.) *Let* $\underline{A} = (A, <)$ *be any order. Suppose* $\underline{B}$ *and* $\underline{C}$ *are both cofinal in A, and* $\operatorname{Ord} B$ *and* $\operatorname{Ord} C$ *are both regular cardinals. Then* $\operatorname{Ord} \underline{B} = \operatorname{Ord} C$. *((II) will be proved below), after* III *and its proof.*

Proposition (III.) (II) $\Rightarrow$ 1.5(c).

The proof of (III) is harder than that of (I) but is straightforward in, say, 5 or 6 'well-chosen' lines.

Proof of (II). We know that $\operatorname{Ord} \underline{B} \le \operatorname{Ord} \underline{C}$ or $\operatorname{Ord} \underline{C} \le \operatorname{Ord} B$. Clearly we can, without loss of generality, assume $\operatorname{Ord} \underline{B} \le \operatorname{Ord} C$. We will obtain (using $\underline{B}$-recursion) an isomorphism f of $\underline{B}$ onto a cofinal subset of $\underline{C}$. We justify this (somewhat informally) as follows: Let $b \in B$, and suppose we have $(f(b') : b' < b)$, an isomorphism of $\{b' : b' < b\}$ onto $\{f(b') : b' < b\} \subseteq C$. Since $\operatorname{Ord} \underline{B}$ is a cardinal, $\{b' : b' < b\}$ and hence $\{f(b') : b' < b\}$ have power $< \overline{\overline{B}} \le \overline{\overline{C}}$, so $\{f(b') : b' < b\}$ is not cofinal in $\underline{C}$, as $\underline{C}$ is regular. Hence we can take c' = the $\underline{C}$-first element $> f(b')$ for all $b' \in B$, and take c'' = the $\underline{C}$-first element $>_{\underline{A}} b$, and $f(b) = \max(c', c'')$. It is easy to see that f is well-defined (by $\underline{B}$-recursion). The reader should prove that $D = f[B]$ is cofinal in C (in 2 lines), so $\overline{\overline{D}} = \overline{\overline{C}}$. Also, he should check that if $\underline{U}$ and $\underline{V}$ any orders such that $\operatorname{Ord} \underline{U}$ and $\operatorname{Ord} \underline{V}$ are regular cardinals and $\overline{\overline{U}} = \overline{\overline{V}}$ then $\operatorname{Ord} U = \operatorname{Ord} V$. So $\operatorname{Ord} C = \operatorname{Ord} B' = \operatorname{Ord} B$.

§2. Problems 1–4. Problems 1–4 already have hints.

Problem 5. Here is a sketch of a proof of 2.5^{AC} with much to be filled in. Theorem 2.5^{AC}: If $\alpha \ge 2$, $\beta \ne 0$, and α is infinite or β is infinite, then $\overline{\overline{\alpha^\beta}} = \max(\overline{\overline{\alpha}}, \overline{\overline{\beta}})$.

Proof. Claim (0) If κ is infinite and $\lambda \ne 0$ then $\max(\kappa, \lambda) \cdot \lambda = \max(\kappa, \lambda)$. (Just take cases $\kappa \le \lambda$ and $\kappa > \lambda$.) Henceforth assume $\alpha \ge 2$ and $\beta \ne 0$.

Claim (1) $\alpha^\beta \ge \beta$. The proof is by induction on $\beta \ge 1$ (with α fixed).

Claim (2) $\alpha^\beta \ge \alpha$. Proof in 2 lines with no induction.

By (1) and (2), $\alpha^\beta \ge \max(\alpha, \beta)$. Hence (3) $\overline{\overline{\alpha^\beta}} \ge \max(\overline{\overline{\alpha}}, \overline{\overline{\beta}})$.

Claim (4) If $\alpha \ge \omega$ then $\overline{\overline{\alpha^\beta}} \le \max(\overline{\overline{\alpha}}, \overline{\overline{\beta}})$.

Do the easy case $0 \ne \beta < \omega$, then prove (4) holds for $\beta \ge \omega$ by induction on $\beta \ge \omega$.

(Note: If $\beta = \delta$, a limit ordinal, then $\overline{\overline{\sup_{\beta<\delta} \alpha^\beta}} \le \sum_{\beta<\delta} \overline{\overline{\alpha^\beta}}$. In the step from β to $\beta + 1$, use (0).)

Claim (5) Take $\alpha = n \ge 2$ and hence β infinite. Then $\overline{\overline{n^\beta}} \le \overline{\overline{\beta}}$ ($= \max(\overline{\overline{n}}, \overline{\overline{\beta}})$).

Proof is by induction on $\beta \ge \omega$. Do $\beta = \omega$ first. Next, assuming that $\beta \ge \omega$ and (5) holds for β, show (5) holds for $\beta + 1$. These two steps are easy. Finally, assume $\beta = \delta$ is a limit ordinal $> \omega$ and $\overline{\overline{n^\gamma}} \le \overline{\overline{\gamma}}$ if $\omega \le \gamma < \delta$. Then $\overline{\overline{n^\delta}} = \overline{\overline{\sup_{\gamma<\delta} n^\gamma}} = \overline{\overline{\sup_{\omega\le\gamma<\delta} n^\gamma}} \le \sum_{\omega\le\gamma<\delta} \overline{\overline{n^\gamma}} = \sum_{\omega\le\gamma\le\delta} \overline{\overline{\gamma}} \le \overline{\overline{\delta}} \cdot \overline{\overline{\delta}} = \overline{\overline{\delta}}$. Thus (5) is proved and 2.5 follows from (3), (4) and (5).

Problem 6. Theorem 2.6(c) *should* read as follows:

> (Representation to the base $\beta \ge 2$.) For any ζ there exists n, $\alpha_0, \ldots, \alpha_{n-1}$, and $\eta_0, \ldots, \eta_{n-1}$ such that $\alpha_0 > \alpha_1 > \cdots > \alpha_{n-1}$ and, for all $i < n$, $0 < \eta_i < \beta$, and $\zeta = \beta^{\alpha_0} \cdot \eta_0 + \beta^{\alpha_1} \cdot \eta_1 + \cdots + \beta^{\alpha_{n-1}} \cdot \eta_{n-1}$; moreover, the representation is unique.

Now the proofs of both existence and uniquenes (in 2.6(c)), using 2.6(a) and (b), are fairly easy.

Problem 7. We shall do (b) $\Rightarrow$ (a) (leaving (a) $\Rightarrow$ (b) as an easy but good problem for the reader).

Note: If $\gamma < \omega^\delta$, then $\gamma < \omega^{\delta'} \cdot p$ for some $\delta' < \delta$, some $p \in \omega$ (by 2.6(b)). Hence, if $\gamma, \gamma' < \omega^\delta$ then for some $\delta' < \omega$, some p, we have $\gamma, \gamma' < a^{\delta'} \cdot p$.

Now assume (b) and suppose (a) fails, *i.e.*, ω^δ can be written as $\gamma + \rho$ where $0 < \rho < \omega^\delta$ (and hence $\gamma < \omega^\delta$). By the Note above we have $\gamma, \rho < \omega^{\delta'} \cdot p$ for some $\delta' < \delta$ and some p. Hence $\omega^\delta = \gamma + \rho \le \omega^{\delta'}(p + p) < \omega^\delta$, which is absurd.

Problem 8. Let $K_{\alpha\beta} = \{f : \beta \to \alpha : f(\gamma) = 0$ for all but finitely many $\gamma < \beta\}$. For *distinct* $f, g \in K_{\alpha\beta}$, put $f \prec g$ iff among the finitely many γ for which $f(\gamma) \ne g(\gamma)$ the largest such γ has $f(\gamma) < g(\gamma)$. Write $K^*_{\alpha\beta}$ for the *structure* $(K_{\alpha,\beta}, \prec)$.

It is clear that $\prec$ is irreflexive and connected. We now show that $\prec$ is transitive. Suppose $f \prec g$ and $g \prec h$. Let $\delta_1 = \max\{\gamma : f(\gamma) \ne g(\gamma)\}$ and $\delta_2 = \max\{\gamma : g(\gamma) \ne h(\gamma)\}$. Consider the three cases: $\delta_1 = \delta_2$, $\delta_1 < \delta_2$, $\delta_2 < \delta_1$. In each case a straightforward argument shows that $f \prec h$. We illustrate with the case $\delta_1 < \delta_2$ and leave the other two cases to the reader. Assume $\delta_1 < \delta_2$. Using the definitions of δ_1 and δ_2, we see the following: Clearly $f(\delta_2) = g(\delta_2)$; but $g(\delta_2) < h(\delta_2)$ so (*) $f(\delta_2) < h(\delta_2)$. Clearly

$f(\xi) = g(\xi) = h(\xi)$ for all ξ such that $\delta_2 \leq \xi < \beta$. Hence, by $(*)$, $f \prec h$, as desired. Thus $K^*_{\alpha\beta}$ is an order.

If $\gamma < \beta$, put $K^\beta_{\alpha\gamma} = \{f \in K_{\alpha\beta}$: for all ξ, if $\gamma \leq \xi < \beta$ then $f(\xi) = 0\}$. Also, let $\Gamma_{\alpha\gamma\beta}$ be the function on $K_{\alpha\gamma}$ such that if $g \in K_{\alpha\gamma}$ then $\Gamma_{\alpha\gamma\beta}(g)$ is the f on β with $f \restriction \gamma = g$ and $f(\xi) = 0$ if $\gamma \leq \xi < \beta$. It is straightforward to check that $K^\beta_{\alpha\gamma}$ is a proper initial segment of $K^*_{\alpha\beta}$ and $\Gamma_{\alpha\gamma\beta}$ is an isomorphism of $K^*_{\alpha\gamma}$ onto $K^{\beta *}_{\alpha\gamma}$ (which is $K^\beta_{\alpha\gamma}$ as a substructure of $K^*_{\alpha\beta}$).

Now we prove by induction on 'β' that, for all β, (!) $K^*_{\alpha\beta}$ has order type α^β (and hence is a well-order). If $\beta = 0$, (!) is clear. Suppose (!) holds for β. If $f \in K_{\alpha,\beta+1}$ then clearly $f \restriction \beta \in K_{\alpha\beta}$ and $f(\beta) < \alpha$. Put $F(f) = (f \restriction \beta, f(\beta))$. We claim $K^*_{\alpha,\beta+1} \underset{F}{\cong} K^*_{\alpha\beta} \cdot \alpha^*$ (where $\cdot$ is the ordinal product of orders and α^* is $(\alpha, \epsilon_\alpha)$. Clearly F is 1-1, on and onto. But we also have $f \prec g$ iff $f(\beta) < g(\beta)$ or ($f(\beta) = g(\beta)$ and $f \restriction \beta \prec g \restriction \beta$), as one easily checks. Thus the claim is proved. Hence $K^*_{\alpha,\beta+1} \cong K^*_{\alpha\beta} \cdot \alpha^* \cong$ (by induction hypothesis) $\alpha^\beta \cdot \alpha = \alpha^{\beta+1}$.

Finally, suppose (!) holds for all $\beta < \delta$ (a limit ordinal). Then for each $\beta < \delta$, ($K^*_{\alpha\beta}$ and hence also) $K^{\delta *}_{\alpha\beta}$ is isomorphic to (α^β, ϵ); let this uniquely determined map be $G_\beta : K^{\delta *}_{\alpha\beta} \to (\alpha^\beta, \epsilon)$. We now have $K_{\alpha,\delta} = \bigcup_{\beta<\delta} K^\delta_{\alpha\beta}$ (where $\beta \mapsto K^\delta_{\alpha\beta}$ is a chain under "is a proper initial segment of"). Also $\alpha^\delta = \bigcup_{\beta<\delta} \alpha^\beta$ (again a chain under "is a proper initial segment of"). Moreover the G_β's are a chain under $\subseteq$. It follows that $G = \bigcup_{\beta<\delta} G_\beta$ is an isomorphism of $K^*_{\alpha\delta}$ onto α^δ. This completes the proof.